FAT CHANCE

In a world where we are constantly being asked to make decisions based on incomplete information, facility with basic probability is an essential skill. This book provides a solid foundation in basic probability theory designed for intellectually curious readers and those new to the subject. Through its conversational tone and careful pacing of mathematical development, the book balances a charming style with informative discussion.

This text will immerse the reader in a mathematical view of the world, giving them a glimpse into what attracts mathematicians to the subject in the first place. Rather than simply writing out and memorizing formulas, the reader will come out with an understanding of what those formulas mean, and how and when to use them. Readers will also encounter settings where probabilistic reasoning does not apply or where our intuition can be misleading. This book establishes simple principles of counting collections and sequences of alternatives and elaborates on these techniques to solve real-world problems both inside and outside the casino. Readers at any level are equipped to consider probability at large and work through exercises on their own.

Benedict Gross is Leverett Professor of Mathematics, Emeritus at Harvard University and Professor of Mathematics at UC San Diego. He has taught mathematics at all levels at Princeton, Brown, Harvard, and UCSD, and served as the Dean of Harvard College from 2003–2007. He is a member of the American Academy of Arts and Sciences and the National Academy of Science. Among his awards and honors are the Cole Prize from the American Mathematical Society and a MacArthur Fellowship. His research is primarily in number theory.

Joe Harris is the Higgins Professor of Mathematics at Harvard University. He has been at Harvard since 1988 and was previously on the faculty at MIT and Brown. He is a member of the American Academy of Arts and Sciences and the National Academy of Science. Throughout his career, he has been deeply committed to education at every level, which led to a partnership with Benedict Gross to develop the Harvard course "Fat Chance," the inspiration for the book of the same title. He is author of several books including *3264 and All That, Algebraic Geometry,* and *The Geometry of Schemes.*

Emily Riehl is an Assistant Professor of Mathematics at Johns Hopkins University and previously was a Benjamin Peirce and NSF postdoctoral fellow at Harvard University. She has published over twenty papers and written two books: *Categorical Homotopy Theory* and *Category Theory in Context.* She has been awarded an NSF grant and a CAREER award to support her work and has been recognized for excellence in teaching at both Johns Hopkins and Harvard.

Fat Chance

Probability from 0 to 1

BENEDICT GROSS

Harvard University

JOE HARRIS

Harvard University

EMILY RIEHL

Johns Hopkins University

CAMBRIDGE
UNIVERSITY PRESS

CAMBRIDGE
UNIVERSITY PRESS

University Printing House, Cambridge CB2 8BS, United Kingdom

One Liberty Plaza, 20th Floor, New York, NY 10006, USA

477 Williamstown Road, Port Melbourne, VIC 3207, Australia

314–321, 3rd Floor, Plot 3, Splendor Forum, Jasola District Centre, New Delhi – 110025, India

79 Anson Road, #06–04/06, Singapore 079906

Cambridge University Press is part of the University of Cambridge.

It furthers the University's mission by disseminating knowledge in the pursuit of
education, learning, and research at the highest international levels of excellence.

www.cambridge.org
Information on this title: www.cambridge.org/9781108482967
DOI: 10.1017/9781108610278

First published 2019

Printed in the United Kingdom by TJ International Ltd., Padstow, Cornwall

A catalogue record for this publication is available from the British Library.

Library of Congress Cataloging-in-Publication Data
Names: Gross, Benedict H., 1950- author. | Harris, Joe, 1951– author. | Riehl, Emily, author.
Title: Fat chance : probability from 0 to 1 / Benedict Gross (Harvard University, Massachusetts),
    Joe Harris (Harvard University, Massachusetts), Emily Riehl (Johns Hopkins University).
Description: Cambridge ; New York, NY : Cambridge University Press, c2019. |
    Includes bibliographical references and index.
Identifiers: LCCN 2018058461| ISBN 9781108482967 (hardback : alk. paper) |
    ISBN 9781108728188 (pbk. : alk. paper)
Subjects: LCSH: Probabilities–Popular works. | Probabilities–Problems, exercises, etc.
Classification: LCC QA273.15 .G76 2019 | DDC 519.2–dc23
    LC record available at https://lccn.loc.gov/2018058461

ISBN 978-1-108-48296-7 Hardback
ISBN 978-1-108-72818-8 Paperback

Contents

Preface

Suppose a friend comes up to you in a bar with the following challenge. He asks you for a quarter—which you hand over somewhat reluctantly—and tells you he's going to flip it six times and record the outcome: heads or tails. Your friend then offers you the following bet: if you can correctly predict the number of heads, he'll buy the next round of drinks. But if the outcome is different from what you guessed, the next round is on you. On each individual flip heads and tails are equally likely. So it seems that three heads and three tails is the most likely outcome. But is this a good bet?

Or, you and a friend are driving to the movies and find a place to park a mile away from the theater. It's a busy area, but you know there's some chance that if you drive on, you'll find a space that's only half a mile away. It's also possible that if you look for a better spot, you'll have to come back to this one, which might no longer be there. How do you decide what to do? And how does it affect your decision if the movie starts in 25 minutes?

In a school election, a vast majority of the left-handed voters prefer Tracy, who has advocated strongly for more left-handed desks to be put in classrooms. Does this mean that Tracy is likely to win? And while we're at it, what does a "3% margin of error" in a poll really mean?

Or suppose you go to the doctor with your uncle and are told that his level of prostate-specific antigen (PSA) is high. The doctor reports that a large proportion of people with prostate cancer have elevated PSA levels. How worried should you be?

Finally, suppose you want to gamble with your life's savings, totaling $1,000, while playing roulette. Is it better to wager it all on a single bet, or to bet just $1 at a time until you either double your money or go bankrupt?

At its heart, probability is about decision-making, at least in contexts when it's possible to determine how mathematically likely a given event would be. In this book we embark upon a journey that will allow us to discover the answers to all of the questions posed above, using the mathematics of probability theory to help us identify the ways in which your intuition may lead you astray.

We'll discuss the gambler's fallacy and its opposite, the hot hand fallacy in basketball, and try to understand why random correlations arise in large data sets, such as the "coincidence" that means that it's likely that two people in a room of 30 share a birthday. We'll explore the mathematical distinction between events that are independent, in the sense of being unrelated, and those that are "positively" or "negatively" correlated. We'll look at games like slots, poker, and roulette to understand how Las Vegas stays in business, even when it's possible to accurately compute the expected value of each game. And we try to intersperse more serious discussions, involving medical decision-making and the zombie apocalypse, with some lighthearted questions we explore just for fun.

Fat Chance began as a course in the program of General Education at Harvard College. Two of us (Dick and Joe) taught it to help students who were not majoring in mathematics or the sciences develop some basic skills in counting and probability. We also hoped to use the course to describe a mathematical view of the world, and to show students what attracted us to math in the first place.

After we gave the course in the classroom a few times, we were approached by HarvardX to develop lectures and problems, in order to present the course online. This book is intended both as a companion to our online course and for the general reader who wants to learn the basics of probability.

Perhaps somewhat counterintuitively, we think of Fat Chance as akin to an introductory language course. The heart of a language course is not the memorization of a lot of vocabulary and verb tenses—though invariably there is a lot of that involved—but rather the experience of thinking and speaking in a different tongue. In the same way, in this book there are of necessity a fair number of techniques to learn and calculations to carry out. But they are just a means to an end: our goal is to give you the experience of thinking mathematically, and the ability to calculate, or at least estimate, probabilities that come up in many real-world situations.

One way to try to learn a new language is to watch a lot of foreign films with the subtitles turned off. Eventually you'll learn to speak a few phrases, but, appealing as this method of studying may sound, it's not exactly the most efficient tactic to achieve fluency. A better strategy would be to find a few native speakers and attempt to have a conversation. You're bound to say the wrong thing on a few occasions, which can be embarrassing, but the evidence suggests you'll learn much faster this way, as much because of your mistakes as despite them. For this reason, we've included some challenge exercises for the reader to puzzle over throughout the text in hopes that many of you will take this opportunity to try out for yourselves the probabilistic reasoning techniques we introduce.

What sort of prerequisites does this book have? Well, the technical answer to that is "very few." Some familiarity with the topics in a high school algebra course (manipulation of fractions, comfort with the use of letters to stand for numbers) should be plenty. Probably more important is a less quantifiable requirement: we ask the reader to be prepared to approach the mathematics with a spirit of adventure and exploration, with the understanding that, while some work will be required, the experience at the end will be well worth it.

We are indebted to a great number of people who helped us create this book. Our treatment of the topics in the first part of the book was clearly influenced by Ivan Niven in his book, *Mathematics of Choice.* Many graduate students at Harvard served as teaching fellows when we gave the course in the classroom, and they helped to shape both the material and the exercises. Special thanks go to Andrew Rawson and all the videographers and editors in the HarvardX studios, who made the filming of Fat Chance such a pleasure, to Cameron Krulewski, who helped teach the online course and gave useful hands-on feedback about the exercises, and to Devlin Mallory, who helped us with copyediting and figure formatting. We are indebted to our editor, Katie Leach, at Cambridge University Press, for her patience and her many helpful suggestions on the possible audience for this book.

Now, let's get down to work!

PART I

Counting

1 Simple counting

It's hard to begin a math book. A few chapters in, it gets easier: by then, writer and reader have—or think they have—a common sense of the level of the book, its pace, language, and goals; at that point, communication naturally flows more smoothly. But getting started is awkward.

As a consequence, it's standard practice in math textbooks to include a throwaway chapter or two at the beginning. These function as a warmup before we get around to the part of the workout that involves the heavy lifting. An introductory chapter typically has little or no technical content, but rather is put there in the hope of establishing basic terminology and notation, and getting the reader used to the style of the book, before launching into the actual material. Unfortunately, the effect may be the opposite: a chapter full of seemingly obvious statements, expressed in vague language, can have the effect of making the reader generally uneasy without actually conveying any useful information.

Well, far be it from us to deviate from standard practice! The following is our introductory chapter. But here's the deal: you can skip it if you find the material too easy (unlike the case of power lifting, where a thorough warmup is absolutely necessary). Really. Just go right ahead to Chapter 2 and start there.

1.1 COUNTING NUMBERS

To start things off, we'd like to talk about counting, because that's how numbers first entered our world. It was four or five thousand years ago that people first developed the concept of numbers, probably in order to quantify their possessions and make transactions—my three pigs for your two cows and the like. And the remarkable thing that people discovered about numbers is that the same system of numbers—1, 2, 3, 4, and so on—could be used to count anything: beads, bushels of grain, people living in a village, forces in an opposing army. Numbers can count anything: numbers can even count numbers.

And that's where we'll start. The first problem we're going to pose is simply: how many numbers are there between 1 and 10?

At this point you may be wondering if it's too late to get your money back for this book. Bear with us! We'll get to stuff you don't know soon enough. In the meantime, write them out and count:

$$1, \quad 2, \quad 3, \quad 4, \quad 5, \quad 6, \quad 7, \quad 8, \quad 9, \quad 10;$$

there are 10. How about between 1 and 11? Well, that's one more, so there are 11. Between 1 and 12? 12, of course.

Well, that seems pretty clear, and if we now asked you, for example, how many numbers there are between 1 and 57, you wouldn't actually have to write them out and count; you'd figure (correctly) that the answer would be 57.

OK, then, let's ramp it up a notch. Suppose we ask now: how many numbers are there between 28 and 83, inclusive? ("Inclusive" means that, as before, we include both 28 and 83 in the count.) Well, you could do this by making a list of the numbers between 28 and 83 and counting them, but you have to believe there's a better way than that.

Here's one: suppose you did write out all the numbers between 28 and 83:

$$28, \quad 29, \quad 30, \quad 31, \quad 32, \quad \ldots\ldots, \quad 82, \quad 83.$$

(Here the dots invite the reader to imagine that we've written all the numbers in between in an unbroken sequence. We'll use this convention when it's not possible or desirable to write out a sequence of numbers in full.) Now subtract the number 27 from each of them. The list now starts at 1, and continues up to $83 - 27 = 56$:

$$1, \quad 2, \quad 3, \quad 4, \quad 5, \quad \ldots\ldots, \quad 55, \quad 56.$$

From what we just saw we know there are 56 numbers on this list; so there were 56 numbers on our original list as well.

It's pretty clear also that we could do this to count any string of numbers. For example, if we asked how many numbers there are between 327 and 573, you could similarly imagine the numbers all written out:

$$327, \quad 328, \quad 329, \quad 330, \quad 331, \quad \ldots\ldots, \quad 572, \quad 573.$$

Next, subtract the number 326 from each of them; we get the list

$$1, \quad 2, \quad 3, \quad 4, \quad 5, \quad \ldots\ldots, \quad 246, \quad 247,$$

and so we conclude that there were $573 - 326 = 247$ numbers on our original list.

Now, there's no need to go through this process every time. It makes more sense to do it once with letters standing for arbitrary numbers, and in that way work out a formula that we can use every time we have such a problem. So, imagine that we're given two whole numbers n and k, with n the larger of the two, and we're asked: how many numbers are there between k and n, inclusive?

We do this just the same way: imagine that we've written out the numbers from k to n in a list

$$k, \quad k+1, \quad k+2, \quad k+3, \quad \ldots\ldots, \quad n-1, \quad n$$

and subtract the number $k - 1$ from each of them to arrive at the list

$$1, \quad 2, \quad 3, \quad 4, \quad \ldots\ldots, \quad n-1-(k-1), \quad n-(k-1).$$

Now we know how many numbers are on the list: it's $n - (k - 1)$ or, more simply, $n - k + 1$.[1] Our conclusion, then, is that:

---

[1]   Is it obvious that $n - (k-1)$ is the same as $n - k + 1$? If not, take a moment out to convince yourself: subtracting $k - 1$ is the same as subtracting k and then adding 1 back. In this book we'll usually carry out operations like this without comment, but you should take the time to satisfy yourself that they make sense.

> The number of whole numbers between k and n inclusive is $n - k + 1$.

So, for example, if someone asked "How many numbers are there between 342 and 576?" we wouldn't have to think it through from scratch: the answer is $576 - 342 + 1$, or 235.

Since this is our first formula, it may be time to bring up the whole issue of the role of formulas in math. As we said, the whole point of having a formula like this is that we shouldn't have to recreate the entire argument we used in the concrete examples above every time we want to solve a similar problem. On the other hand, it's also important to keep some understanding of the process, and not to treat the formula as a "black box" that spews out answers (regardless of the black box we drew to call attention to the general formula). Knowing how the formula was arrived at helps us to know both when it's applicable, and how it can be modified to deal with other situations when it's not.

Exercise 1.1.1. How many whole numbers are there between 242 and 783?

Exercise 1.1.2. The collection of whole numbers, together with their negatives and the number zero, form a number system called the *integers*.

1. Suppose n and k are both negative numbers. How many negative numbers are there between k and n inclusive?
2. Suppose n is positive and k is negative. How many integers are there between k and n inclusive?

1.2 COUNTING DIVISIBLE NUMBERS

Now that we've done that, let's try a slightly different problem: suppose we ask "How many even numbers are there between 46 and 104?"

In fact, we can approach this the same way: imagine that we did make a list of all even numbers, starting with 46 and ending with 104:

$$46, \quad 48, \quad 50, \quad 52, \quad \ldots\ldots, \quad 102, \quad 104.$$

Now, we've just learned how to count numbers in an unbroken sequence. And we can convert this list to just such a sequence if we just divide all the numbers on the list by 2: doing that, we get the sequence

$$23, \quad 24, \quad 25, \quad 26, \quad \ldots\ldots, \quad 51, \quad 52$$

of all whole numbers between $46/2$, or 23, and $104/2$, or 52. Now, we know by the formula we just worked out how many numbers there are on that list: there are

$$52 - 23 + 1 \; = \; 30$$

numbers between 23 and 52, so we conclude that there are 30 even numbers between 46 and 104.

One more example of this type: let's ask the question, "How many numbers between 50 and 218 are divisible by 3?" Once more we use the same approach: imagine that we made a list of all such numbers. But notice that 50 isn't the first such number, since 3 doesn't divide 50 evenly: in fact, the smallest number on our list that is divisible by 3 is $51 = 3 \times 17$. Likewise, the last number on our list is 218, which isn't divisible by 3. The largest number on our list which is divisible by 3 is 216, which is: $3 \times 72 = 216$. So the list of numbers divisible by 3 would look like

$$51, \quad 54, \quad 57, \quad 60, \quad \ldots\ldots, \quad 213, \quad 216.$$

Now we can do as we did before, and divide each number on this list by 3. We arrive at the list

$$17, \quad 18, \quad 19, \quad 20, \quad \ldots\ldots, \quad 71, \quad 72$$

of all whole numbers between 17 and 72, and there are

$$72 - 17 + 1 \;=\; 56$$

such numbers.

Now it's time to stop reading for a moment and do some yourself:

Exercise 1.2.1.

1. How many numbers between 242 and 783 are divisible by 6?
2. How many numbers between 17 and 783 are divisible by 6?
3. How many numbers between 45 and 93 are divisible by 4?

Exercise 1.2.2.

1. In a sports stadium with numbered seats, every seat is occupied except seats 33 through 97. How many seats are still available?
2. Suppose the fans are superstitious and only want to sit in even-numbered seats because otherwise they fear that their team will lose. How many even-numbered seats are still available in the block of seats numbered 33 through 97?

Exercise 1.2.3. In a non-leap year of 365 days starting on Sunday, January 1st, how many Sundays will there be? How many Mondays will there be?

1.3 "I'VE REDUCED IT TO A SOLVED PROBLEM."

Note one thing about the sequence of problems we've just done. We started with a pretty mindless one—the number of numbers between 1 and n—which we could answer more or less by direct examination. The next problem we took up—the number of numbers between k and n—we solved by shifting all the numbers down to whole numbers between 1 and $n - k + 1$. In effect we reduced it to the first problem, whose answer we knew. Finally, when we asked how many numbers between two numbers were divisible by a third, we answered the question by dividing all the numbers, to reduce the problem to counting numbers between k and n.

This approach—building up our capacity to solve problems by reducing new problems to ones we've already solved—is absolutely characteristic of mathematics. We start out slowly, and gradually accumulate a body of knowledge and techniques; the goal is not necessarily to solve each problem directly, but to reduce it to a previously solved problem.

There's even a standard joke about this:

A mathematician walks into a room. In one corner, she sees an empty bucket. In a second corner, she sees a sink with a water faucet. And, in a third corner, she sees a pile of papers on fire. She leaps into action: she picks up the bucket, fills it up at the faucet, and promptly douses the fire.

The next day, the same mathematician returns to the room. Once more, she sees a fire in the third corner, but this time sitting next to it there's a full bucket of water. Once more she leaps into action: she picks up the bucket, drains it into the sink, places it empty in the first corner and leaves, announcing: "I've reduced it to a previously solved problem!"

Well, maybe you had to be there. But there is a real point to be made here. It's simply this: the ideas and techniques developed in this book are cumulative, each one resting on the foundation of the ones that have come before. We'll occasionally go off on tangents and pursue ideas that won't be used in what follows, and we'll try to tell you when that occurs. But for the most part, *you need to keep up*: that is, you need to work with the ideas and techniques in each section until you feel genuinely comfortable with them, before you go on to the next.

It's worth remarking also that the cumulative nature of mathematics in some ways sets it apart from other fields of science. The theories of physics, chemistry, biology, and medicine we subscribe to today flatly contradict those held in the seventeenth and eighteenth centuries—it's fair to say that medical texts dealing with the proper application of leeches are of interest primarily to historians, and we'd bet your high school chemistry course didn't cover phlogiston.[2] By contrast, the mathematics developed at that time is the cornerstone of what we're doing today.

Exercise 1.3.1. How many numbers between 242 and 783 are *not* divisible by 6?

Exercise 1.3.2. A radio station mistakenly promises to give away two concert tickets to *every* thirteenth caller as opposed to offering two concert tickets only to *the* thirteenth caller. They receive 428 calls before the station manager realizes the mistake. How many concert tickets has the radio station promised to give away?

1.4   REALLY BIG NUMBERS

As long as we're talking about the origins of numbers, let's talk about another important early development: the capacity to write down really big numbers. Think about it: once you've developed the concept of numbers, the next step is to figure out a way to write them down. Of course, you can just make up an arbitrary new symbol for each new number, but this is inherently limited: you can't express large numbers without a cumbersome dictionary.

One of the first treatises ever written on the subject of numbers and counting was by Archimedes, who lived in Syracuse (part of what was then the Greek empire) in the third century BC. The paper, entitled *The Sand Reckoner*, was addressed to a local monarch, and in it Archimedes claimed that he had developed a system of numbers

---

[2] In case you're curious, phlogiston was the hypothetical principle of fire, of which every combustible substance was in part composed—at least until the whole theory was discredited by Antoine Lavoisier between 1770 and 1790.

that would allow him to express as large a number as the number of grains of sand in the universe—a revolutionary idea at the time.

What Archimedes had developed was similar to what we would call *exponential notation*. We'll try to illustrate this by expressing a really large number—say, the approximate number of seconds in the lifetime of the universe.

The calculation is simple enough. There are 60 seconds in a minute, and 60 minutes in an hour, so the number of seconds in an hour is

$$60 \times 60 \ = \ 3{,}600.$$

There are in turn 24 hours in a day, so the number of seconds in a day is

$$3{,}600 \times 24 \ = \ 86{,}400;$$

and since there are 365 days in a (non-leap) year, the number of seconds in a year is

$$86{,}400 \times 365 \ = \ 31{,}536{,}000.$$

Now, in exponential notation, we would say this number is roughly 3 times 10 to the 7th power—that is, a three with seven zeros after it. Here 10^7 refers to the product $10 \times 10 \times 10 \times 10 \times 10 \times 10 \times 10$ of 10 with itself seven times. In standard decimal notation, $10^7 = 10{,}000{,}000$, a one with seven zeros after it, and thus $3 \times 10^7 = 30{,}000{,}000$ is a three with seven zeros after it. (A better approximation, of course, would be to say the number is roughly 3.1×10^7, or 3.15×10^7, but we're going to go with the simpler estimate 3×10^7.)

Exponential notation is particularly convenient when it comes to multiplying large numbers. Suppose, for example, that we have to multiply $10^6 \times 10^7$. Well, 10^6 is just $10 \times 10 \times 10 \times 10 \times 10 \times 10$, and 10^7 is just $10 \times 10 \times 10 \times 10 \times 10 \times 10 \times 10$, so when we multiply them we just get the product of 10 with itself 13 times: that is,

$$10^6 \times 10^7 \ = \ 10^{13}.$$

In other words, we simply add the exponents. So it's easy to take products of quantities that you've expressed in exponential notation.

For example, to take the next step in our problem, we have to say how old the universe is. Now, this quantity very much depends on your model of the universe. Most astrophysicists estimate that the universe is approximately 13.7 billion years old, with a possible error on the order of 1%. We'll write the age of the universe, accordingly, as

$$13{,}700{,}000{,}000 \ = \ 1.37 \times 10^{10}$$

years. So the number of seconds in the lifetime of the universe would be approximately

$$(1.37 \times 10^{10}) \times (3 \times 10^7) \ = \ 4.11 \times 10^{17};$$

or, rounding it off, the universe is roughly 4×10^{17} seconds old.

You see how we can use this notation to express arbitrarily large numbers. For example, computers currently can carry out on the order of 10^{12} operations a second (a *teraflop*, as it's known in the trade). We could ask: if such a computer were running from the dawn of time to the present, how many operations could it have performed? The answer is, approximately,

$$10^{12} \times (4 \times 10^{17}) = 4 \times 10^{29}.$$

Now, for almost all of this book, we'll be dealing with much much smaller numbers than these, and we'll be doing exact calculations rather than approximations. But occasionally we will want to express and estimate larger numbers like these. (The last number above—the number of operations a computer running for the lifetime of the universe could perform—will actually arise later on in this book: we'll encounter mathematical processes that require more than this number of operations to carry out.) But even if there aren't enough seconds since the dawn of time to carry out such a calculation, it's nice to know that we have a notation that can accommodate it.

Exercise 1.4.1. We computed the approximate age of the universe—roughly 4×10^{17} seconds old—back in 2002, so our calculation is several years old. Update our work to compute the approximate age of the universe in seconds as of today.

Exercise 1.4.2. The Library of Alexandria is estimated to have held as many as 400,000 books (really papyrus scrolls), while the US Library of Congress currently holds about 2.8×10^6 books. How many volumes is this in total?

1.5   IT COULD BE WORSE

Look: this is a math book. We're trying to pretend it isn't, but it is. That means that it'll have jargon—we'll try to keep it to a minimum, but we can't altogether avoid using technical terms. That means that you'll encounter the odd mathematical formula here and there. That means it'll have long discussions aimed at solving artificially posed problems, subject to seemingly arbitrary hypotheses. Mathematics texts have a pretty bad reputation, and we're sorry to say it's largely deserved.

Just remember: it could be worse. You could, for example, be reading a book on Kant. Now, Immanuel Kant is a towering figure in Western philosophy, a pioneering genius who shaped much of modern thought. "The foremost thinker of the Enlightenment and one of the greatest philosophers of all time," the Encyclopedia Britannica calls him. But just read a sentence of his writing:

> If we wish to discern whether anything is beautiful or not, we do not refer the representation of it to the Object by means of understanding with a view to cognition, but by means of the imagination (acting perhaps in conjunction with understanding) we refer the representation to the Subject and its feeling of pleasure or displeasure.

What's more, this is not a nugget unearthed from deep within one of Kant's books. It is, in fact, the first sentence of the first Part of the first Moment of the first Book of the first Section of Part I of Kant's *The Critique of Judgement*.

Now, we're not trying to be anti-intellectual here, or to take cheap shots at other disciplines. Just the opposite, in fact: what we're trying to say is that any body of thought, once it progresses past the level of bumper-sticker catchphrases, requires a language and a set of conventions of its own. These provide the precision and universality that are essential if people are to communicate and develop the ideas further, and shape them into a coherent whole. But they also can have the unfortunate effect of making much of the material inaccessible to a casual reader. Mathematics suffers from this—as do most serious academic disciplines.

The point, in other words, is not that the passage from Kant we just quoted is babble; it's not. (Lord knows we could have dug up enough specimens of academic writing that are, if that was our intention.) In fact, it's the beginning of a serious and extremely influential attempt to establish a philosophical theory of aesthetics. As such, it may be difficult to understand without some mental effort. It's important to bear in mind that the apparent obscurity of the language is a reflection of this difficulty, not necessarily the cause of it.

So, the next time you're reading this book and you encounter a term that turns out to have been defined—contrary to its apparent meaning—some 30 pages earlier, or a formula that seems to come out of nowhere and that you're apparently expected to find self-explanatory, just remember: it could be worse.

2 The multiplication principle

2.1 CHOICES

Let's suppose you climb out of bed one morning, still somewhat groggy from the night before. You grope your way to your closet, where you discover that your cache of clean clothes has been reduced to four shirts and three pairs of pants. It's far too early to exercise any aesthetic judgment whatsoever: any shirt will go with any pants; you only need something that will get you as far as the cafe around the corner and that blessed, life-giving cup of coffee. The question is:

> How many different outfits can you make out of your four shirts and three pairs of pants?

Admittedly the narrative took a sharp turn toward the bizarre with that last sentence. Why on earth would you or anyone care how many outfits you can make? Well, bear with us while we try to answer it anyway.

Actually, if you thought about the question at all, you probably have already figured out the answer: each of the four shirts is part of exactly three outfits, depending on which pants you choose to go with it, so the total number of possible outfits is $3 \times 4 = 12$. (Or, if you like to get dressed from the bottom up, each of the three pairs of pants is part of exactly four outfits; either way the answer is 3×4.) If we're feeling really fussy, we could make a table: say the four shirts are a polo shirt, a button-down, a tank top, and a T-shirt extolling the virtues of your favorite athletic wear, and the pants consist of a pair of jeans, some cargo pants, and a pair of shorts. Then we can arrange the outfits in a rectangle:

polo shirt & jeans	button-down & jeans	tank top & jeans	T-shirt & jeans
polo shirt & cargo pants	button-down & cargo pants	tank top & cargo pants	T-shirt & cargo pants
polo shirt & shorts	button-down & shorts	tank top & shorts	T-shirt & shorts

Now, you know we're not going to stop here. Suppose next that, in addition to picking a shirt and a pair of pants, you also have to choose between two pairs of shoes. Now how many outfits are there?

Well, the idea is pretty much the same: for each of the possible shirt/pants combinations, there are two choices for the shoes, so the total number of outfits is

$4 \times 3 \times 2 = 12 \times 2 = 24$. And if in addition we had a choice of five hats, the total number of possible outfits would be $4 \times 3 \times 2 \times 5 = 120$—you get the idea.

Now it's midday and you head over to the House of Pizza to order a pizza for lunch. You feel like having one meat topping and one vegetable topping on your pizza; the House of Pizza offers you seven meat toppings and four vegetable toppings. How many different pizzas do you have to choose among?

"That's the same problem with different numbers!" you might say, and you'd be right: to each of the seven meat toppings you could add any one of the four vegetable toppings, so the total number of different pizzas you could order would be 7×4, or 28.

Evening draws on, and your roommates ask you to queue up a triple feature to watch. They request one action film, one romantic comedy, and one comedy special. The film streaming service you subscribe to has 674 action films (most of which were direct-to-video), 913 romantic comedies (ditto), and 84 comedy specials. How many triple features can you watch?

"That's the same problem again!" you might be thinking: the answer's just the number 674 of action movies times the number 913 of romantic comedies times the number 84 of comedy specials, or

$$674 \times 913 \times 84 = 51{,}690{,}408.$$

Trust us—we are going somewhere with this. But you're right, it's time to state the general rule that we're working toward, which is called the *multiplication principle*:

> The number of ways of making a sequence of independent choices is the product of the number of choices at each step.

Here "independent" means that how you make the first choice doesn't affect the number of choices you have for the second, and so on. In the first case above, for example—getting dressed in the morning—it corresponds to having no fashion sense whatsoever.

The multiplication principle is easy to understand and apply, but awkward to state in reasonably coherent English, which is why we went through three examples before announcing it. In fact, you may find the examples more instructive than the principle itself; if the boxed statement seems obscure to you, just remember: "4 shirts and 3 pants equals 12 outfits."

Exercise 2.1.1. You have seven colors of nail polish and 10 fingers. You're not coordinated enough to use more than one color on each nail, but aren't fussed about using different colors on different nails. How many ways can you paint the nails on your hands if you are not worried about whether they match?

2.2   COUNTING WORDS

An old-style license plate has on it a sequence of three numbers followed by three letters. How many different old-style license plates can there be?

This is easy enough to answer: we have 10 choices for each of the numbers, and 26 choices for each of the letters; and since none of these choices is constrained in any way, the total number of possible license plates is

$$10 \times 10 \times 10 \times 26 \times 26 \times 26 \ = \ 17{,}576{,}000.$$

A similar question is this: Suppose for the moment that by "word" we mean any finite sequence of the 26 letters of the English alphabet—we're not going to make a distinction between actual words and arbitrary sequences. How many three-letter words are there?

This is just the same as the license plate problem (or at least the second half): we have 26 independent choices for each of the letters, so the number of three-letter words is $26^3 = 17{,}576$. In general,

$$
\begin{aligned}
\text{\# of 1-letter words} &= 26, \\
\text{\# of 2-letter words} &= 26^2 = 676, \\
\text{\# of 3-letter words} &= 26^3 = 17{,}576, \\
\text{\# of 4-letter words} &= 26^4 = 456{,}976, \\
\text{\# of 5-letter words} &= 26^5 = 11{,}881{,}376, \\
\text{\# of 6-letter words} &= 26^6 = 308{,}915{,}776,
\end{aligned}
$$

and so on.

Next, let's suppose that there are 15 students in a class, and that they've decided to choose a set of class officers: a president, a vice president, a secretary and a treasurer. How many possible slates are there? That is, how many ways are there of choosing the four officers?

Actually, there are two versions of this question, depending on whether or not a single student is allowed to hold more than one of the positions. If we assume first that there's no restriction of how many positions one person can hold, the problem is identical to the ones we've just been looking at: we have choices each for the four offices, and they are all independent, so that the total number of possible choices is

$$15 \times 15 \times 15 \times 15 \ = \ 50{,}625.$$

Now suppose on the other hand that we impose the rule that no person can hold more than one office. How many ways are there of choosing officers now?

Well, this can also be computed by the multiplication principle. We start by choosing the president; we have clearly 15 choices there. Next, we choose the vice president. Now our choice is restricted by the fact that our newly selected president is no longer eligible, so that we have to choose among the 14 remaining students. After that we choose a secretary, who could be anyone in the class except the two officers already chosen, so we have 13 choices here; finally we choose a treasurer from among the 12 students in the class other than the president, vice president, and secretary. Altogether, the number of choices is

$$15 \times 14 \times 13 \times 12 \ = \ 32{,}760.$$

Note one point here: in this example, the actual choice of, say, the vice-president *does* depend on who we chose for president; the choice of a secretary does depend on who we selected for president and vice president, and so on. But the *number* of choices doesn't depend on our prior selections, so the multiplication principle still applies.

In a similar vein, we could modify the question we asked a moment ago about the number of three-letter words, and ask: how many three-letter words have no repeated letters? The solution is completely analogous to the class-officer problem: we have 26 choices for the first letter, 25 for the second, and 24 for the third, so that we have a total of

$$26 \times 25 \times 24 \ = \ 15{,}600$$

such words.

In general, we can calculate

$$\text{\# of 1-letter words} \ = \ 26,$$
$$\text{\# of 2-letter words without repeated letters} \ = \ 26 \cdot 25 \ = \ 650,$$
$$\text{\# of 3-letter words without repeated letters} \ = \ 26 \cdot 25 \cdot 24 \ = \ 15{,}600,$$
$$\text{\# of 4-letter words without repeated letters} \ = \ 26 \cdot 25 \cdot 24 \cdot 23 \ = \ 358{,}800,$$
$$\text{\# of 5-letter words without repeated letters} \ = \ 26 \cdot 25 \cdot 24 \cdot 23 \cdot 22$$
$$= \ 7{,}893{,}600,$$
$$\text{\# of 6-letter words without repeated letters} \ = \ 26 \cdot 25 \cdot 24 \cdot 23 \cdot 22 \cdot 21$$
$$= \ 165{,}765{,}600,$$

and so on. Note that here we are using a simple dot $\cdot$ in place of the times symbol $\times$. In general, when we have an expression with a lot of products, we'll use this simpler notation to avoid clutter. Sometimes we'll omit the product sign altogether; for example, we write $2n$ for $2 \times n$.

Now, here's an interesting (if somewhat tangential) question. Let's compare the numbers of words of each length to the number of words with no repeated letters. What percentage of all words have repetitions, and what percentage doesn't? Of course, as the length of the word increases, we'd expect a higher proportion of all words to have repeated letters—relatively few words of two or three letters have repetitions, while necessarily every word of 27 or more letters does. We could ask, then: when does the fraction of words without repeated letters dip below one-half? In other words, for what lengths do the words with repeated letters outnumber those without?

Before we tabulate the data and give the answer, you might want to take a few minutes and think about the question. What would your guess be?

length	number of words	without repeats	% without repeats
1	26	26	100.00
2	676	650	96.15
3	17,576	15,600	88.76
4	456,976	358,800	78.52
5	11,881,376	7,893,600	66.44
6	308,915,776	165,765,600	53.66
7	8,031,810,176	3,315,312,000	41.28
8	208,827,064,576	62,990,928,000	30.16
9	5,429,503,678,976	1,133,836,704,000	20.88

Now that's bound to be surprising: among six-letter words, those with repeated letters represent nearly half, and among seven-letter words they already substantially outnumber the words without repeats. In general, the percentage of words without repeated letters drops off pretty fast: by the time we get to twelve-letter words, fewer than 1 in 20 has no repeated letter. We'll see another example of this phenomenon when we talk about the birthday problem in Section 5.6.

Exercise 2.2.1. Suppose you want to select a class president, vice president, secretary, and treasurer, under the condition that no person can hold more than one office, but this time the plan is to choose the treasurer first, then the secretary, then the vice president, and then the president. How many ways are there of choosing the four officers now? How does this relate to the computation performed above?

Exercise 2.2.2. The Greek alphabet has 24 letters while the Russian alphabet consists of 33 letters.

1. Predict whether the percentage of words of length n in the Greek alphabet without repeats will be smaller or greater than the percentage of words of length n in the English alphabet without repeats.
2. Predict whether the percentage of words of length n in the Russian alphabet without repeats will be smaller or greater than the percentage of words of length n in the English alphabet without repeats.
3. Compute the number of words of length 1, 2, 3, 4, and 5 with and without repeats in Greek and in Russian and see whether or not your predictions were correct.

2.3   A SEQUENCES OF CHOICES

There are two special cases of the multiplication principle that occur so commonly in counting problems that they're worth mentioning on their own, and we'll do that here. Neither will be new to us; we've already encountered examples of each.

Both involve a sequence of selections from a single pool of objects. If there are no restrictions at all on the choices, the application of the multiplication principle is particularly simple: each choice in the sequence is a choice among all the objects in the collection. If we're counting three-letter words in an alphabet of 26 characters, for example—where by "word" we again mean an arbitrary sequence of letters—there are 26^3; if we're counting four-letter words in an alphabet of 22 characters, there are 22^4; and so on. In general, we have the following rule:

> The number of sequences of k objects chosen from a collection of n objects is n^k.

The second special case involves the same problem, but with a commonly applied restriction: we're again looking at sequences of objects chosen from a common pool of objects, but this time we're not allowed to choose the same object twice. Thus, the first choice is among all the objects in the pool; the second choice is among all but one, the third among all but two, and so on; if we're looking at a sequence of k choices, the last choice will be among all but the $k - 1$ already chosen. Thus, as we saw, the number of three-letter words without repeated letters in an alphabet of 26 characters is $26 \cdot 25 \cdot 24$; the number of four-letter words without repeated letters in an alphabet of

22 characters is $22 \cdot 21 \cdot 20 \cdot 19$; and so on. In general, if the number of objects in our pool is n, the first choice will be among all n, the second among $n - 1$, and so on. If we're making a total of k choices, the last choice will exclude the $k - 1$ already chosen; that is, it'll be a choice among the $n - (k - 1) = n - k + 1$ objects remaining. The total number of such sequences is thus the product of the numbers from n down to $n - k + 1$. We write this as

$$n \cdot (n - 1) \cdot (n - 2) \cdot \cdots \cdot (n - k + 1),$$

where the dots in the middle indicate that you're supposed to keep going multiplying all the whole numbers in the series starting with n, $n - 1$ and $n - 2$, until you get down to $n - k + 1$. Time for a box:

The number of sequences of k objects chosen without repetition from a collection of n objects is $n \cdot (n - 1) \cdot (n - 2) \cdot \cdots \cdot (n - k + 1)$.

Exercise 2.3.1. In the state lottery, the winning number is chosen by picking six ping-pong balls from a bin containing balls labeled "1" through "36" to arrive at a sequence of six numbers between 1 and 36. Ping-pong balls are not replaced after they're chosen; that is, no number can appear twice in the sequence. How many possible outcomes are there?

Note that in this last exercise, the order in which the ping-pong balls are chosen is relevant: if the winning sequence is "17-32-5-19-12-27" and you picked "32-17-5-19-12-27," you *don't* get to go to work the next day and tell your boss what you really think of her.

Exercise 2.3.2. The Hebrew alphabet has 22 letters. How many five-letter words are possible in Hebrew? (Again, by "word" we mean just an arbitrary sequence of five characters from the Hebrew alphabet.) What fraction of these have no repeated letters?

2.4   FACTORIALS

The two formulas we described in the last section were both special cases of the multiplication principle. There is in turn a special case of the second formula that crops up fairly often and that's worth talking about now. We'll start, as usual, with an example.

Problem 2.4.1. Suppose that we have a first-grade class of 15 students, and we want to line them up to go out to recess. How many ways of lining them up are there—that is, in how many different orders can they be lined up?

Solution. Well, think of it this way: we have 15 choices of who'll be first in line. Once we've chosen the line leader, we have 14 choices for who's going to be second, 13 choices for the third, and so on. In fact, all we're doing here is choosing a sequence of 15 children from among the 15 children in the class, without repetition; whether we invoke the formula in the last section or do it directly, the answer is

$$15 \cdot 14 \cdot 13 \cdot 12 \cdot 11 \cdot 10 \cdot 9 \cdot 8 \cdot 7 \cdot 6 \cdot 5 \cdot 4 \cdot 3 \cdot 2 \cdot 1 = 1{,}307{,}674{,}368{,}000,$$

or about 1.3×10^{12}—more than a trillion orderings. $\square$

In general, if we ask how many ways there are of placing n objects in a sequence, the answer is the product of all the whole numbers between 1 and n. This is a quantity that occurs so often in mathematics (and especially in counting problems) that it has its own symbol and name:

> The product $n \cdot (n-1) \cdot (n-2) \cdots \cdot 3 \cdot 2 \cdot 1$ of the numbers from 1 to n is written $n!$ and called "n factorial."

Here's a table of the factorials up to 15:

n	$n!$
1	1
2	2
3	6
4	24
5	120
6	720
7	5,040
8	40,320
9	362,880
10	3,628,800
11	39,916,800
12	479,001,600
13	6,227,020,800
14	87,178,291,200
15	1,307,674,368,000

There are many fascinating things to be said about these numbers. Their size alone is an interesting question: we've seen that 15 factorial is over a trillion; approximately how large a number is, say, 100 factorial? But we'll leave these questions aside for now. At this point, we'll be using factorials for the most part just as a way of simplifying notation. We'll start with the last formula of the preceding section.

It's pretty obvious that writing 15! is a whole lot easier than writing out the product $15 \cdot 14 \cdot 13 \cdot 12 \cdot 11 \cdot 10 \cdot 9 \cdot 8 \cdot 7 \cdot 6 \cdot 5 \cdot 4 \cdot 3 \cdot 2 \cdot 1$. But there are other, less obvious uses of the notation. Suppose, for example, that we wanted to make up a baseball team out of the 15 kids in the class—that is, choose a sequence of nine of the kids in the class of 15, without repetition. We'd have 15 choices for the pitcher, 14 for the catcher, 13 for the first baseman, and so on. When you choose the ninth and last player, you'll be choosing among the $15 - 8 = 7$ kids left at that point, so that the total number of teams would be

$$15 \cdot 14 \cdot 13 \cdot 12 \cdot 11 \cdot 10 \cdot 9 \cdot 8 \cdot 7.$$

But there's a faster way to write this number, using factorials. Basically, we could think of this product as the product of all the numbers from 15 down to 1, except we leave off the numbers from 6 down to 1—in other words, the product of the numbers from 15 to 1 divided by the product of the numbers from 6 to 1, or

$$\frac{15!}{6!}.$$

Now, this may seem like a strange way of writing out the product: it seems inefficient to multiply all the numbers from 15 to 1 and then divide by the product of the numbers you didn't want in the first place. And it is—no one in their right mind would calculate out the number that way. But just as notation, "$15!/6!$" takes up a whole lot less space than "$15 \cdot 14 \cdot 13 \cdot 12 \cdot 11 \cdot 10 \cdot 9 \cdot 8 \cdot 7$," and we'll go with it. For example, we'll rewrite the boxed formula from the last section:

The number of sequences of k objects chosen without repetition from a collection of n objects is
$$\frac{n!}{(n-k)!}.$$

One final note about factorial notation: it is the standard convention that $0! = 1$. You could think of this as the answer to the Zen koan, "How many ways are there of ordering no objects?" But we'll ignore the philosophical ramifications and simply accept it as a notational convention: it just makes the formulas come out better, as we'll see.

Exercise 2.4.2. For $0 < j < k < n$ which is larger: $n!/j!$ or $n!/k!$? Why?

Exercise 2.4.3. Which is larger: $n!$ or n^n? Why?

Exercise 2.4.4. Thirteen athletes from around the world are competing in the steeplechase competition at the 2020 Olympics. By the *outcome* of the event we'll mean the determination of who gets the gold medal, who gets the silver medal, and who gets the bronze medal. How many possible outcomes are there?

2.5   WHEN ORDER MATTERS

The multiplication principle itself is completely straightforward. But sometimes there may be more than one way to apply it, and sometimes one of those ways will work when another doesn't. We have to be prepared, in other words, to be flexible in applying the multiplication principle. We'll see lots of examples of this over the course of this part of the book; here's one of them.

To start with, let's take a simple problem: how many three-digit numbers can you form using the digits 1 through 9, with no repeated digit? As we've seen already, this is completely straightforward: we have nine choices for the first digit, then eight choices for the second, and finally seven choices for the third, for a total of

$$9 \times 8 \times 7 = 504$$

choices.

Now let's change the problem a bit: suppose we ask, "How many of those 504 numbers are odd?" In other words, how many have as their third digit a 1, 3, 5, 7, or 9?

We can try to do it the same way: as before, there are nine choices for the first digit and eight for the second. But when we get to the third digit, we're stuck. For example, if the first two digits we selected were 2 and 4, then the third digit could be any of the numbers 1, 3, 5, 7, or 9, so we'd have five choices. If the first two digits were 5 and 7, however, the third digit could only be a 1, 3, or 9; we'd have only three choices. The choices, in other words, don't seem to be independent.

But they are if we make them in a different order! Suppose that rather than choosing the first digit first and so on, we go from right to left instead—in other words, choose the third digit first, then the middle, and finally the first. Now we can choose the last digit freely among the numbers 1, 3, 5, 7, and 9, for a total of five choices. The choice of the middle digit is constrained only by the requirement that it not repeat the one we've already chosen; so, there are eight choices for it, and likewise seven choices for the first digit. There are thus

$$5 \times 8 \times 7 = 280$$

such numbers.

Sometimes we find ourselves in situations where the multiplication principle may not seem applicable, but in fact its application is completely straightforward as long as we keep our wits about us. Here's an example:

Problem 2.5.1. The pupils in a class of 15 students must wear their school uniforms—a polo shirt that's either orange or black—and on this day eight pupils have opted for a orange polo while the other seven are wearing a black polo. Suppose that for aesthetic reasons the teacher wants to line up the students so that no two orange-shirted students are standing next to each other. How many ways are there of doing this?

Solution. Actually, before we go and give the solution, let's take a moment and see that the multiplication principle fails. In fact, if we try to use the same approach as we took to in solving Problem 2.4.1, it screws up already at the second step. That is, we seemingly have as before 15 choices of who's to be first in line. But the numbers of possible choices for who goes second depends on our first choice: if we chose a black-shirted student to be first, there are no restrictions on who goes second, and there are 14 choices; but if we chose a orange-shirted student to be first in line, the second in line must be chosen from among the seven students wearing black.

We need, in other words, a different approach. But here we're in luck: if we think about it, we can see that since there are 8 orange shirts out of 15 kids, and no two orange-shirted students are to be next to each other in line, the line must alternate orange/black/orange/black until the final place, which must be a student wearing orange. In other words, the odd-numbered places in line must all be occupied by orange-shirted students, and the even places by the black-shirted ones.

Thus, to choose an ordering of the whole class subject to the constraint that no two orange-shirted students are next to each other, we have to choose an ordering for the eight orange-shirted students and separately choose an ordering for the seven black-shirted students, and then line up the whole class together starting with the first orange-shirted student and then alternating shirt colors from there. We know that there are 8! ways of ordering eight students and 7! ways of ordering seven students, so the multiplication principle tells us that the total number of ways of lining up the class is

$$8! \cdot 7! = 203{,}212{,}800. \qquad \square$$

Here's a related puzzler:

Problem 2.5.2. Suppose that there were six students wearing orange and nine students wearing black, and again we wanted to line up the class so that no two orange-shirted students are next to each other. How many ways would there be of doing this?

In fact, this is a *much* harder problem, because we can't avail ourselves of the trick we used in Problem 2.5.1. But it is one you'll learn how to do. So think for a while about how you might try to approach it, and we promise we'll work it out in Section 4.3.

Exercise 2.5.3.

1. How many numbers are there between 100 and 999 with no repeated digits?
2. How many three-digit odd numbers are there?

Exercise 2.5.4. Consider all numbers consisting of four different digits all between 1 and 9. How many of these are odd?

3 The subtraction principle

There are 26 letters in the English alphabet, divided into vowels and consonants. For the purposes of this discussion, we'll say the letters

$$A, E, I, O, \text{ and } U$$

are vowels, and the letters

$$B, C, D, F, G, H, J, K L, M, N, P, Q, R, S, T, V, W, X, Y, \text{ and } Z$$

are consonants. Now, quickly: *how many consonants are there?*

How many of you counted out the letters in the sequence B, C, D, F, . . .? Probably not many: it's just a lot easier to count the vowels, and subtract the number of vowels (5) from the total number of letters (26) to arrive at the answer that there are $26 - 5 = 21$ consonants.

And that's all there is to the subtraction principle, which is the second of the basic counting tools we'll be using, after the multiplication principle. It's not at all deep—it amounts to nothing more than an observation, really—but we'll give it a box anyway:

> The number of objects in a collection that satisfy some condition is equal to the total number of objects in the collection minus the number of those that don't.

The point being, it's often easier to count the latter than the former. It hardly warrants a box of its own, but—in conjunction with the multiplication principle—it gives us a number of different ways of approaching a lot of counting problems. In fact, as we'll see in this chapter and the next, it greatly broadens the scope of the problems we can solve.

We start with a few simple examples:

3.1 COUNTING THE COMPLEMENT

Once more, you're going to have a triple feature in your room: one action film, one romantic comedy and one comedy special. The catalog of your streaming services lists 674 action films (489 of which feature car chases), 913 romantic comedies (217 of which feature car chases), and 84 comedy specials (two of which feature a car chase, inexplicably). But there's one restriction: some of your roommates have informed you that if you select three movies featuring car chases they're officially kicking you out of the room. Now how many triple bills are possible?

Well, we could try to do this with the multiplication principle, as we did before the anti-car-chase faction in your room raised its voice. But it's easy to see this isn't going to work. We can pick the action movie freely, of course; we have 674 choices there. And we can pick the romantic comedy freely as well; that's a free choice among 913 movies. But when it comes time to pick the last movie, how many choices we have depends on what our choices up to that point have been: if either of the first two movies is without car chases, we can choose the third movie freely among the 84 comedy specials; but if both of our first two choices do feature car chases, the choice of the third movie is limited to those 82 that don't. Changing the order of selection doesn't help, either: any way we work it, the number of choices available to us for the last movie depends on our first two selections.

So what do we do? It's simple enough. We already know how many total choices we'd have if there were no restrictions: as we worked it out, it's just

$$674 \times 913 \times 84 = 51{,}690{,}408.$$

At the same time, it's easy enough to count the number of triple bills that are excluded if we want to stay in the room: we can choose any of the 489 action movies featuring car chases, any of the 217 romantic comedies featuring car chases, and any of the two comedy specials featuring car chases, for a total of

$$489 \times 217 \times 2 = 212{,}226$$

disallowed triple features. The number of allowable choices is thus

$$51{,}690{,}408 - 212{,}226 \ = \ 51{,}478{,}182,$$

which should be plenty.

Here's a similar problem (some might say the same problem). We've already counted the number of four-letter words, by which we mean arbitrary sequences of four of the 26 characters in the English alphabet. Suppose we ask now, *how many such words have at least one vowel*? (Here we'll stick to the convention that "Y" isn't a vowel.)

As in the last problem, the multiplication principle seems to work fine until we get to the last letter, and then it breaks down. We have 26 choices for the first letter, 26 for the second, and 26 for the third. But when it comes to choosing the last letter, we don't know how many choices we'll have: if any of the preceding three choices happened to be a vowel, we are now free to choose any letter for the last one in our word; but if none of the first three was a vowel we can only choose among the five vowels for the last.

Instead, we use the subtraction principle: we know how many words there are altogether, and *we'll subtract from that the number of words consisting entirely of consonants*. Both are easy: the number of all possible words is just 26^4, and the number of four-letter words consisting only of consonants is 21^4, so the answer to our problem is

$$26^4 - 21^4 \ = \ 456{,}976 - 194{,}481 \ = \ 262{,}495.$$

One more example: in the first section, we saw how to answer questions like, "How many numbers are there between 34 and 78?" and "How many numbers between 34 and 78 are divisible by 5?" Well, suppose now someone asks, "How many numbers between 34 and 78 are *not* divisible by 5?"

It's pretty clear this is a case for the subtraction principle. We know the number of numbers between 34 and 78 is

$$78 - 34 + 1 = 45.$$

Moreover, since the first and last numbers between 34 and 78 that are divisible by 5 are $35 = 7 \times 5$ and $75 = 15 \times 5$, the number of numbers in this range divisible by 5 is the number of numbers between 7 and 15; that is,

$$15 - 7 + 1 = 9.$$

So by the subtraction principle, the number of numbers between 34 and 78 that are not divisible by 5 is $45 - 9$, or 36.

Exercise 3.1.1. Your boss, Ebenezer Scrooge, grudgingly grants you Saturdays and Sundays off work, but insists that you come into the office every Monday, Tuesday, Wednesday, Thursday, and Friday, regardless of any public holidays.

1. How many days must you work in a non-leap year beginning on a Sunday?
2. How many days must you work in a non-leap year beginning on a Tuesday?
3. Is this arrangement fair?

Exercise 3.1.2. Continuing Exercise 2.4.4, say that 3 of the 13 athletes in the steeplechase competition are from Moldova (steeplechase is big in Moldova, apparently). How many outcomes—the determination of who gets the gold, who gets the silver, and who gets the bronze—involve at least one Moldovan winning a medal?

Exercise 3.1.3. How many six-letter words in the English alphabet have at least one repeated letter?

Exercise 3.1.4. Let's assume that a phone number has seven digits, and cannot start with a 0.

1. How many possible phone numbers are there?
2. How many phone numbers are there with at least one even digit?

3.2 THE ART OF COUNTING

With the subtraction principle, we've doubled the number of techniques we can apply to counting problems. One downside to having more than one technique, though, is that it's no longer unambiguous how to go about solving a problem: we may need to use one technique, or the other, or a combination. This is the beginning of the art of counting, and to develop our technique we'll work out a few more examples.

For a start, let's return to our fifteen-student classroom, but this time without worrying what color shirt each student is wearing. This time, though, let's make the problem a little more difficult: let's suppose that two of the kids in the class, Becky and Ethan, are truly obnoxious little brats. Either one individually is unruly to the point of psychopathy; the last thing in the world you'd want is the two of them standing next to each other in line. So, the problem we're going to deal with is:

Problem 3.2.1. How many ways are there of lining up the class so Becky and Ethan are *not* next to each other?

Solution. Well, we can certainly apply the subtraction principle here: we know there are 15! ways there are of ordering the class if we pose no restrictions; so, if we can figure out how many ways there are of lining them up so that Becky and Ethan *are* next to each other, we can subtract that from the total and get the answer that way.

So, how do we figure out the number of lineups with Becky and Ethan adjacent? It seems we haven't exactly solved the problem yet: the next thing we see is that the multiplication principle isn't going to work here, at least not as we applied it in Problem 2.4.1. We can certainly choose any of the 15 students to occupy the first place in line, but then the number of choices for the second place in line depends on whether the first choice was Becky or Ethan, or one of the other 13 kids. What's more, this ambiguity persists at every stage thereafter: whom we can put in each place in line depends on whom we put in the preceding spot.

But there are other ways of applying the multiplication principle in this setting. In the solution we gave to Problem 2.4.1, we made our choices one place at a time—that is, we chose one of the 15 kids to occupy the first place in line; then we chose one of the remaining 14 kids to occupy the second place, and so on. But we could have done it the other way around: we could have taken the kids one at a time, and assigned each a place in line. For example, we could start with Becky, and assign her any of the 15 places in line; then, go on to Ethan and assign him any of the remaining 14 places in line, and so on through all 15 kids.

As long as we're dealing with the version of the problem given in Problem 2.4.1, it doesn't matter which approach we take; both lead us to the answer $15 \cdot 14 \cdot 13 \cdots 3 \cdot 2 \cdot 1 = 15!$. But in the current situation—where we're trying to count the number of lineups with Becky and Ethan adjacent, say—it does make a difference.

At first it may not seem like it. Doing it this way, we can assign Becky to any of the 15 places in line, but then the number of choices we have for Ethan depends on where we assigned Becky: if Becky was placed in either the first or the fifteenth place in line, we will have no choice but to place Ethan in the second or fourteenth, respectively; but if Becky was placed in any of the interior slots, then we can choose to place Ethan either immediately ahead of her or immediately behind her. So it seems that the multiplication principle doesn't work this way, either.

But there is a difference. Approaching the problem this way—taking the students one at a time, and assigning each in turn one of the remaining places in line—we see that once we've got Becky and Ethan assigned to their places, the multiplication principle takes over: there are 13 choices for where to place the next kid, 12 choices of where to place the one after that, and so on. In other words, if we break the problem up into first assigning Becky and Ethan their places, and then assigning the remaining 13 kids theirs, we see that

$$\left\{ \begin{array}{l} \text{the number of lineups} \\ \text{of the class with Becky} \\ \text{and Ethan adjacent} \end{array} \right\} = \left\{ \begin{array}{l} \text{the number of ways of} \\ \text{assigning Becky and Ethan} \\ \text{adjacent places in line} \end{array} \right\} \times 13! \, .$$

It remains to count the number of ways of assigning Becky and Ethan adjacent places in line. This is not hard: as we saw above, there are two ways of doing this with Becky occupying an end position, and $13 \times 2 = 26$ ways of doing it with Becky occupying an interior position (second through fourteenth), for a total of 28 ways. Or we could count this way: to specify adjacent places in line for Becky and Ethan, we could first specify the *pair* of positions they're to occupy—first and second, or second and third, and so on up to fourteenth and fifteenth—and then say which of the pair Becky is to occupy. For the first, there are 14 choices, and for the latter 2 choices, so by the multiplication principle we see again there are 28 ways of assigning Becky and Ethan adjacent places in line.

In conclusion, we see that

$$\left\{ \begin{array}{l} \text{the number of lineups} \\ \text{of the class with Becky} \\ \text{and Ethan adjacent} \end{array} \right\} = 28 \times 13!$$

and, correspondingly,

$$\left\{ \begin{array}{l} \text{the number of lineups} \\ \text{of the class with Becky} \\ \text{and Ethan apart} \end{array} \right\} = 15! - (28 \times 13!)$$

$$= 1{,}133{,}317{,}785{,}600. \qquad \square$$

Exercise 3.2.2. Do Problem 3.2.1 over, using the approach followed above but without the subtraction principle: that is, count the number of lineups of the class with Becky and Ethan apart by counting the number of ways you can assign Becky and Ethan to two *nonadjacent* places in line, and the number of ways you can assign the remaining 13 students to the remaining 13 places. Does your answer agree with the one above?

Before we go on, we want to emphasize one point that is illustrated by Problem 3.2.1 and its solution. It's an important aspect of learning and doing mathematics, and the failure to appreciate it is the cause of a lot of the frustration that everyone experiences in reading math books. Simply put, it's this: *formulas don't work.* At least, they don't usually work in the sense that you can just plug in appropriate numbers, turn the crank and arrive at an answer. It's better to think of formulas as guides, suggesting effective ways of thinking about problems.

That's probably not what you wanted to hear. When it's late at night and your math homework is the only thing standing between you and bed, you don't want to embark on a glorious journey of exploration and discovery. You just want someone to tell you what to do to get the answer, and formulas may appear to do exactly that. But, really, that's not what they're there for, and appreciating that fact will spare you a lot of aggravation.

Now you try it.

Exercise 3.2.3. A new-style license plate has two letters (which can be any letter from A to Z) followed by four numbers (which can be any digits from 0 to 9).

1. How many new-style license plates are there?
2. How many new-style license plates are there if we require no repeated letters and no repeated numbers?
3. How many new-style license plates are there that have at least one "7"?

Exercise 3.2.4. Getting dressed: suppose you own eight shirts, five pairs of pants, and three pairs of shoes.

1. Assuming you have no fashion sense whatsoever, how many outfits can you make?
2. Suppose now that you can make any combination *except* ones including the red pants and the purple shirt. How many outfits can you make?
3. Now suppose that any time you wear the purple shirt you *must* also wear the red pants. How many outfits can you make?

The following problem is hard, but doesn't use any ideas that we haven't introduced.

Exercise 3.2.5. Let's go back to the problem of lining up our class of 15 students. Suppose that Becky and Ethan are so wired that for the sake of everyone's sanity we feel there should be at least two other kids between them. Now how many possible lineups are there?

3.3 MULTIPLE SUBTRACTIONS

Even as simple an idea as the subtraction principle sometimes has complications. In this section, we'll discuss some of what can happen when we have to exclude more than one class of object from a pool. As with the subtraction principle itself, the basic concept is more common sense than arithmetic, and to emphasize that point we'll start with an edible example.

Consider the following list of 17 vegetables:

artichokes
asparagus
beets
broccoli
cabbages
carrots
cauliflower
celery
corn
eggplant
lettuce
onions
peas
peppers
potatoes
spinach
zucchini

Of these, four—beets, carrots, onions, and potatoes—are root vegetables. Two—corn and potatoes—are starchy. Now we ask the question: how many are neither root vegetables nor starchy?

Well, the obvious thing to do would be to subtract the number of root vegetables and starchy vegetables from the total, getting the answer

$$17 - 4 - 2 = 11.$$

But a moment's thought (or, for that matter, actual counting) shows you that isn't right: because a potato is both a root vegetable and a starchy one, you've subtracted it twice. The correct answer is accordingly 12.

And that's the point of this section. It amounts to the observation that when you want to exclude two classes of objects from a pool and count the number left, you can start with the total number of objects in the pool and subtract the number of objects in each of the two excluded categories; *but then you have to add back in the number of objects that belong to both classes and have therefore been subtracted twice.*

The reason has to do with something mathematicians call the *inclusion-exclusion principle*: if A and B are sets of objects in a pool—such as the sets of root vegetables and starchy vegetables considered above—the number of elements in their union is equal to the number of elements in A plus the number of elements in B minus the number of elements in their intersection. If we write $A \cup B$ for the union of A and B and $A \cap B$ for the intersection of A and B, the inclusion-exclusion principle can be expressed concisely as follows:

> For any two sets of elements in a given pool, the number of elements in their union is equal to the sum of the number of elements in each set minus the number of elements in their intersection:
> $$\#(A \cup B) = \#A + \#B - \#(A \cap B).$$

So if you want to *exclude* both A and B from a count of objects, you have to subtract the number of elements in A and subtract the number of elements in B, but then add back in the number of elements in their intersection.

Here's a more mathematical example:

Problem 3.3.1. How many numbers between 100 and 1,000 are divisible by neither 2 nor 3?

Solution. We know that the number of numbers between 100 and 1,000 is simply

$$1,000 - 100 + 1 = 901.$$

Likewise, we can count the numbers in this range divisible by 2: these are just the even numbers between 100 and 1,000, or in other words twice the numbers between 50 and 500; so there are

$$500 - 50 + 1 = 451$$

of them. Similarly, the numbers divisible by 3 are just 3 times the numbers between 34 and 333; so there are

$$333 - 34 + 1 = 300$$

of those. So, naively, we want to subtract each of 451 and 300 from the total 901.

But, as you've probably figured out—seeing as we've stepped all over this punchline —that would be wrong. Because there are numbers divisible by both 2 and 3, and these will have been subtracted twice; to rectify the count we have to add them back in once.

Now, what numbers are divisible by both 2 and 3? The answer is that a number divisible by 2 and by 3 is necessarily divisible by 6, and vice versa.[1] So, such numbers between 100 and 1,000 are just the numbers in that range divisible by 6, which is to say 6 times the numbers between 17 and 167. There are thus

$$167 - 17 + 1 \; = \; 151$$

of them, and so the correct answer to our problem will be

$$901 - 451 - 300 + 151 \; = \; 301. \qquad \square$$

Here's one more involved example of the same idea. Again, we're keeping the convention that by a "word" we mean an arbitrary sequence of letters of the English alphabet.

Problem 3.3.2. How many four-letter words are there in which no letter appears three or more times in a row?

Solution. This clearly calls for the subtraction principle. We know how many four-letter words there are in all—the number is

$$26 \times 26 \times 26 \times 26 \; = \; 456{,}976.$$

We just have to subtract the number of words in which a letter appears three or more times in a row.

Now, there are two kinds of four-letter words in which a letter appears three times in a row: those in which the first three letters are the same, and those where the last three letters are the same. In each case, the number of such words is easy to count by the multiplication principle. For example, to specify a word in which the first three letters are the same, we have to specify that letter (26 choices) and the last letter (26 choices again), so there are

$$26 \times 26 \; = \; 676$$

of this type. By the same token, there are 676 four-letter words in which the last three letters are the same; so naively we want to exclude $2 \times 676 = 1{,}352$ words.

But once more that's not quite right: the 26 words in which all four letters are the same belong to both classes, and so have been subtracted twice! So to correct the count, we have to add them back in once. The correct answer is therefore

$$456{,}976 - 1{,}352 + 26 \; = \; 455{,}650.$$

Actually, there's another way to do this that amounts to the same calculation but avoids the issue of multiple subtractions. We can count the number of words in which one letter appears *exactly* three times in a row, and the number of words in which one letter appears four times, add them up and subtract the total from the number of all four-letter words. For the first, there are again two classes of such words, but within each class the number is different: we choose the repeated letter among the 26

---

[1] This has to do with the *fundamental theorem of arithmetic*, which says that every whole number factors uniquely into a product of prime numbers. Since $6 = 2 \cdot 3$, it follows that a number is divisible by 6 if and only if it is divisible by both 2 and 3.

letters of the alphabet as before, but since that letter is to appear exactly three times the remaining letter must be chosen among the remaining 25 letters of the alphabet. There are thus a total of

$$2 \times 26 \times 25 \;=\; 1{,}300$$

such words. There are again 26 words in which one letter appears all four times; so the correct answer is

$$456{,}976 - 1{,}300 - 26 \;=\; 455{,}650,$$

as before.   □

This last exercise represents another level of complexity in the subtraction principle, but you should be able to do it if you keep your wits about you.

Exercise 3.3.3. How many five-letter words are there in which no letter appears three or more times in a row?

Exercise 3.3.4. A phone number has seven digits and cannot begin with zero.

1. How many phone numbers are there?
2. How many phone numbers contain at least one 7?
3. How many phone numbers contain the sequence 123?

Exercise 3.3.5. Generalize the inclusion-exclusion principle to three sets:

1. If A, B, and C are three sets of elements in a given pool, give a formula for the number of elements in their union $A \cup B \cup C$.
2. Use the formula determined above to figure out how to count the number of elements in the pool that are *not* in A or in B or in C.

4 Collections

4.1 COLLECTIONS VS. SEQUENCES

In this chapter we're going to introduce a new, fundamental idea in counting. This will also be the last new formula: using this and the ideas we've already introduced in combination, we'll be able to count all the objects we want, at least until the final (and optional) chapter of this part of the book.

There's nothing mysterious about it. Basically, in the last couple of chapters we've considered a range of problems in which we count the number of ways to make a sequence of choices. In each instance, either the choices were made from different collections of objects (shirts and pants; meat and vegetable toppings on our pizza; action thrillers and light romantic comedies) or, if they were selections made from the same collection of objects, the order mattered: when we're counting four-letter words, "POOL" is not the same as "POLO."

What we want to look at now are situations where we choose a collection of objects from the same pool, and *the order doesn't matter*. We'll start by revisiting some of the problems we've dealt with, and show how slight variations will put us in this kind of situation.

4.1 COLLECTIONS VS. SEQUENCES

It's a new day, and once more you head over to the House of Pizza for lunch. Today, though, you're feeling both hungry and carnivorous: a pizza with three meat toppings sounds about right. Assuming that the House of Pizza is still offering seven meat toppings, how many different pizzas will fit the bill?

On to the library. Your uncle asked you to pick up four picture books about dinosaurs to entertain your niece. The library has 23 picture books with dinosaurs in stock. How many different choices do you have at the library?

Finally, let's consider a high school class of 15 students. This time, we're not going to select officers; we're just going to choose a committee of four students. There are no distinctions among the four members of the committee; we just have to select four students from among the 15 in the class. How many different committees can be formed?

You get the idea? In each case we're choosing a collection of a specified number of objects from a common pool (toppings, books, students), and *the order doesn't matter*: ordering a pizza with sausage, pepperoni, and hamburger gets you pretty much the same pizza as ordering one with hamburger, pepperoni, and sausage. This sort of situation comes up constantly: when you're dealt a hand of five cards in draw poker, or of

13 cards in bridge, it doesn't matter in what order you receive the cards; possible hands consist of collections of five or 13 cards out of the deck of 52. By way of language, in this sort of situation—where we make a series of selections from a common pool, but all that matters is the totality of objects selected, not the order in which they're selected—we'll refer to choosing a *collection* of objects. In settings where the selections are made from different pools, or where the order does matter, we'll refer to choosing a *sequence*.

Now, as you'll recognize, all of the problems above are really the same problem with different numbers substituted. In fact, there are only two numbers involved: in each of these cases, the number of possible choices depends really only on the number of objects in the pool we're selecting from, and the number of objects to be selected to form our collection.

What we need to do, then, is to find a formula for the number of such collections. We'll do that in the following section, and then we'll see how to combine that formula with the others we've derived to solve a large range of counting problems.

4.2   BINOMIAL COEFFICIENTS

The good news: the formula for the number of collections is very simple to write down and to remember. The bad: it's not quite as straightforward to derive as the ones we've done up to now; in fact, figuring it out requires a somewhat indirect argument. What we'll do is show how to find the answer in a particular case, and once we've done that it'll be pretty clear how to replace the particular numbers in that example with arbitrary ones.

Let's take the case of choosing a committee of four students from among a class of 15—that is, the problem of counting the number of possible committees that can be formed. Again, this is a situation where the order of selection doesn't matter: egos aside, choosing Trevon and then Sofía has the same outcome as choosing Sofía and then Trevon; all that matters in the end is who is on the committee and who is not. And since possible committees don't correspond to sequences of choices, the multiplication principle doesn't seem to apply.

But it does apply—in a sort of weird, backhanded way. To see how, let's focus for a moment on a different problem, the class officer problem: that is, counting the number of ways we can choose a president, vice president, secretary, and treasurer for the class, assuming no student can occupy more than one office. As we saw, the multiplication principle works just fine here: we choose a president (15 choices), then the vice president (14), then the secretary (13), and finally the treasurer (12), for a total of

$$15 \cdot 14 \cdot 13 \cdot 12, \qquad \text{or} \qquad \frac{15!}{11!},$$

possible slates.

But now suppose we want to solve the same problem in a different, somewhat warped way (though again using the multiplication principle). Suppose that instead of choosing the slate one officer at a time, we break the process up into two steps: first we choose a committee of four students who will be the class officers, and *then* choose which of those four will be president, vice president, secretary, and treasurer.

That may seem like an unnecessarily complicated way to proceed. After all, we already know the answer to the class officer problem, while we don't know the number of committees. Bear with us! Let's look anyway at what it tells us.

The one thing we do know is, having selected the four members of the committee, how many ways there are of assigning to the four of them the jobs of president, vice president, secretary, and treasurer: by what we already know, this is just $4 \cdot 3 \cdot 2 \cdot 1 = 4! = 24$. So if we do break up the process of selecting class officers into two stages, choosing a committee and then assigning them the four jobs, what the multiplication principle tells us is that

$$\begin{Bmatrix} \text{the number of ways of} \\ \text{choosing a committee} \end{Bmatrix} \cdot 4! = \begin{Bmatrix} \text{the number of ways of} \\ \text{selecting class officers} \end{Bmatrix} = \frac{15!}{11!}.$$

Now that, if you think about it a moment, tells us something. Since we know that the number of ways of selecting class officers is $15!/11!$, we can solve this equation for the number of committees:

$$\begin{aligned} \begin{Bmatrix} \text{the number of ways of} \\ \text{choosing a committee} \end{Bmatrix} &= \frac{1}{4!} \cdot \begin{Bmatrix} \text{the number of ways of} \\ \text{selecting class officers} \end{Bmatrix} \\ &= \frac{15!}{4!\,11!}. \end{aligned}$$

In English: since every choice of committee corresponds to $4! = 24$ different possible choices of class officers, the number of possible committees is simply $(1/24)^{\text{th}}$ the number of slates.

You can probably see from this that it's going to be the same when we count the number of ways of choosing a collection of any number k of objects from a pool of any number n. We know that the number of ways of choosing a *sequence* of k objects without repetition—a first, then a second different from the first, then a third different from the first two, and so on—is just

$$n \cdot (n-1) \cdots (n-k+1) = \frac{n!}{(n-k)!}.$$

At the same time, for each possible collection of k objects from the pool, there are

$$k \cdot (k-1) \cdots 2 \cdot 1 = k!$$

ways of putting them in order—choosing a first, a second, and so on. The conclusion, then, is that

$$\begin{Bmatrix} \text{the number of ways of choos-} \\ \text{ing a collection of } k \text{ objects} \\ \text{without repetition from a pool} \\ \text{of } n \text{ objects} \end{Bmatrix} = \frac{1}{k!} \cdot \begin{Bmatrix} \text{the number of ways of choosing} \\ \text{a sequence of } k \text{ objects with-} \\ \text{out repetition from a pool of } n \\ \text{objects} \end{Bmatrix}$$

$$= \frac{n!}{k!\,(n-k)!},$$

or, in other words:

> The number of ways of choosing a collection of k objects, without repetition, from among n objects is $\dfrac{n!}{k!\,(n-k)!}$.

So, for example, if the House of Pizza offers seven meat toppings, the number of possible pizzas you can order with three meat toppings is

$$\frac{7!}{3!\,4!} = \frac{5{,}040}{6 \cdot 24} = 35;$$

and if you're setting up a streaming queue with instructions to choose an assortment of exactly four of their 23 movies based on comic strips or video games, your choice is among

$$\frac{23!}{4!\,19!} = 8{,}855$$

such assortments.

The numbers that appear in this setting are so ubiquitous in math that they have a name and a notation of their own. They're called *binomial coefficients* (for reasons we'll explain in Chapter 6), and written in this way:

$$\binom{n}{k} = \frac{n!}{k!\,(n-k)!}.$$

There are a number of things we can say right off the bat about binomial coefficients.

To begin with, there is the basic observation that

$$\binom{n}{k} = \binom{n}{n-k}.$$

This is obvious from the above formula: we see that

$$\frac{n!}{k!\,(n-k)!} = \frac{n!}{(n-k)!\,k!}$$

just by rearranging the factors $k!$ and $(n-k)!$ in the denominator. It's also clear from the interpretation of these numbers: after all, specifying which four kids in the class of 15 are to be put on the committee is the same as to specifying which 11 to leave off it; and in general choosing which k objects to take from a pool of n is the same as choosing which $n-k$ not to take.

Second, as we've pointed out, the standard formula for the binomial coefficients

$$\binom{n}{k} = \frac{n!}{k!\,(n-k)!}$$
$$= \frac{n \cdot (n-1) \cdot (n-2) \cdot \cdots \cdot 2 \cdot 1}{k \cdot (k-1) \cdot \cdots \cdot 2 \cdot 1 \cdot (n-k) \cdot (n-k-1) \cdot \cdots \cdot 2 \cdot 1}$$

is in some ways not the most efficient way to represent the number—it's certainly not how you would calculate it in practice—since there are factors that appear in both the numerator and the denominator, and can be canceled. Doing this gives us two alternative ways of writing the binomial coefficient:

$$\binom{n}{k} = \frac{n \cdot (n-1) \cdots \cdots (n-k+1)}{k \cdot (k-1) \cdots \cdots 2 \cdot 1}$$
$$= \frac{n \cdot (n-1) \cdots \cdots (k+1)}{(n-k) \cdot (n-k-1) \cdots \cdots 2 \cdot 1}.$$

This is not just an aesthetic issue, it's a practical one as well. Suppose, for example, you wanted to count the number of possible five-card hands from a standard deck of 52—that is, you wanted to evaluate the binomial coefficient $\binom{52}{5}$—and you wanted to carry out the calculation on your calculator. If you write the binomial coefficient as

$$\binom{52}{5} = \frac{52 \cdot 51 \cdot 50 \cdot 49 \cdot 48}{5 \cdot 4 \cdot 3 \cdot 2 \cdot 1},$$

your calculator will have no trouble multiplying and dividing out the factors. But if you write

$$\binom{52}{5} = \frac{52!}{5!\,47!}$$

you're in trouble: when you punch in 52! in your calculator you'll probably get an error message; most calculators can't handle numbers that large. Or, even worse, you won't get an error message; your calculator will simply switch to scientific notation. In effect, the calculator will round off the number and not tell you; and these round-off errors can and often do become significant.

Before we go on, let's look at some special cases of binomial coefficients. To begin with, note that for any n,

$$\binom{n}{1} = n,$$

corresponding to the statement that "there are n ways of choosing one object from among n." (Well, that's not exactly news.) Also, by our convention that $0! = 1$, we see that

$$\binom{n}{0} = \frac{n!}{0!\,n!} = 1$$

and similarly $\binom{n}{n} = 1$. Again, think of this simply as a convention; it makes the various formulas we're going to discover in Chapter 6 work.

The first interesting case is the number of ways of choosing a pair of objects from among n:

$$\binom{n}{2} = \frac{n(n-1)}{2},$$

so that the number of ways of choosing two objects from among three is $3 \cdot 2/2 = 3$ (remember, this is the same as the number of ways of picking one object); the number of ways of choosing two objects from among four is $4 \cdot 3/2 = 6$; and in general we can make a table

n	# of ways of choosing two objects from among n
3	3
4	6
5	10
6	15
7	21
8	28

and so on. We can make a similar table for the binomial coefficients $\binom{n}{3}$:

n	# of ways of choosing three objects from among n
4	4
5	10
6	20
7	35
8	56
9	84

Mathematicians have found many fascinating patterns in these numbers, as well as other interpretations of them. We'll take a look at a few of these in Chapter 6.

There's one final remark we want to make about the binomial coefficients. From the formula

$$\binom{n}{k} = \frac{n!}{k!\,(n-k)!}$$

it's clear that $\binom{n}{k}$ is a fraction, but it's far from clear that it's actually a whole number. Of course, we know it's a whole number from the interpretation as the number of ways of choosing k objects from n, but that just raises the question: can we see *why* the formula above always yields a whole number? In some cases we can do this. For example, when we look at the formula

$$\binom{n}{2} = \frac{n(n-1)}{2}$$

and ask, "why is this a whole number?" we have an answer: no matter what n is, either n or $n-1$ must be even. Thus the product $n(n-1)$—the numerator of our fraction—must be even, and so the quotient is a whole number. Likewise, consider the formula

$$\binom{n}{3} = \frac{n(n-1)(n-2)}{6}.$$

Of the three factors n, $n-1$ and $n-2$ that appear in the numerator, at least one must be divisible by 3, and at least one must be even. The numerator must thus be divisible by 6, and so the quotient is whole.

But it gets less and less obvious as k increases. For example, when we say that

$$\binom{n}{4} = \frac{n(n-1)(n-2)(n-3)}{24}.$$

is a whole number, we're saying in effect that *the product of any four whole numbers in a row is divisible by* 24. Think about this for a moment: could you convince yourself of this fact without using the interpretation of $\binom{n}{4}$? Could you convince someone else?

Now it's time for you to do some exercises.

Exercise 4.2.1.

1. Suppose now that the menu at the House of Pizza lists eight meat toppings. How many different pizzas can you order there with two (different) meat toppings? How many with three?
2. If the menu at the House of Pizza also lists four vegetable toppings, how many pizzas can be ordered with two meat and two vegetable toppings?

Exercise 4.2.2. Suppose you are given an exam with 10 problems, and are asked to do exactly seven of them. In how many ways can you choose which seven to do?

Exercise 4.2.3. Which is larger, $\binom{n}{k}$ or $\binom{n}{n-k}$? Why?

4.3   COUNTING COLLECTIONS

We can combine the formula we have now for the number of collections with other formulas and techniques. Here are some examples, in the form of solved problems.

Problem 4.3.1. Suppose once more we're asked to choose four students from a high school class of 15 to form a committee, but this time we have a restriction: we don't want the committee to consist of all juniors or all seniors. Suppose there are eight seniors and seven juniors in the class. How many different committees can we form?

Solution. This clearly calls for the subtraction principle. We know that the total number of possible committees is

$$\binom{15}{4} = \frac{15 \cdot 14 \cdot 13 \cdot 12}{4 \cdot 3 \cdot 2 \cdot 1} = 1,365.$$

The number of committees consisting of all seniors is similarly

$$\binom{8}{4} = \frac{8 \cdot 7 \cdot 6 \cdot 5}{4 \cdot 3 \cdot 2 \cdot 1} = 70,$$

and the number of committees consisting of all juniors is

$$\binom{7}{4} = \frac{7 \cdot 6 \cdot 5 \cdot 4}{4 \cdot 3 \cdot 2 \cdot 1} = 35.$$

If we exclude those, we see that the number of allowable committees is

$$\binom{15}{4} - \binom{8}{4} - \binom{7}{4} = 1,365 - 70 - 35 = 1,260.$$

Note that even though this is a multiple subtraction, we don't need to add any terms back in, since there are no committees that belong to both excluded classes—that is, a committee can't simultaneously consist of all seniors and all juniors. □

Problem 4.3.2. One more committee: now suppose we require that the committee includes exactly two seniors and two juniors. How many possibilities are there now?

Solution. This, by contrast, is a case for the multiplication principle: to choose an allowable committee subject to this restriction, we simply have to choose two among the eight seniors, and (independently) two among the seven juniors. The answer is thus

$$\binom{8}{2} \cdot \binom{7}{2} = 28 \cdot 21 = 588.$$ □

Problem 4.3.3. There are 10 players on a basketball team, and the coach is going to divide them up into two teams of five—the Red team and the Blue team, say—for a practice scrimmage. She's going to do it randomly, meaning that all of the $\binom{10}{5}$ ways of assigning the players to the two teams are equally likely.

Two of the players, Elena and Sofía, are friends. "I hope we wind up on the same team," Sofía says. "Well, we have a 50-50 chance," Elena replies.

Is Elena right?

Solution. What this problem is asking us to do is to count, among the $\binom{10}{5}$ ways of assigning players to the two teams, how many result in Elena and Sofía being on the same team, and how many result in their being on opposing teams. Elena is, in effect, saying that these two numbers will be equal; we'll calculate both and see if she's right.

Let's start by counting the number of ways of choosing the teams that result in Elena and Sofía winding up on the same team. We can specify such an assignment in two stages: first, we decide which team, Red or Blue, gets the Sofía/Elena duo. Obviously, there are two possibilities. Having done that, we have to take the remaining eight players and divide them into two groups: three will go to the team that already has Elena and Sofía; five will go the other team. The number of ways of doing that is $\binom{8}{3}$, so the total number of team assignments with the two as teammates is

$$2 \cdot \binom{8}{3} = 2 \cdot 56 = 112.$$

Now let's count the number of assignments that result in Elena and Sofía opposing each other. Again, we can specify such a choice in two steps: first, we specify which team, Red or Blue, Sofía is on; Elena will necessarily go the other. We then have to take the remaining eight players and assign four of them to each of the two teams; there are $\binom{8}{4}$ ways of doing this, and so the number of team assignments with Sofía and Elena opposite is

$$2 \cdot \binom{8}{4} = 2 \cdot 70 = 140.$$

The conclusion, then, is that Elena is wrong: it is more likely that she and Sofía will wind up on opposing teams.

We're not done! One thing we should always look for in doing these problems is a way to check the accuracy both of our analysis and of our calculations. Here we have a perfect way to do that. We've said that there are a total of $\binom{10}{5}$ ways of assigning the 10 players to the two teams, of which 112 result in the two friends being teammates and 140 result in their being on opposite teams. Before we're satisfied that we've got the correct answer, we should check that in fact $\binom{10}{5}$ is equal to the sum $112 + 140 = 252$. Let's do it:

$$\binom{10}{5} = \frac{10 \cdot 9 \cdot 8 \cdot 7 \cdot 6}{5 \cdot 4 \cdot 3 \cdot 2 \cdot 1}$$
$$= \frac{30,240}{120}$$
$$= 252.$$

Having done this, we can be much more confident both that our analysis was correct, and that we didn't make any numerical mistakes. We can also say that the probability that Sofía and Elena wind up on the same team is

$$\frac{112}{252} = \frac{4}{9} \sim .444$$

or about 44%. □

Problem 4.3.4. Let's suppose you're playing liars Scrabble—a game which requires you to rearrange the letters on your rack into "words" that may or may not be found in an English dictionary. In your rack, you have the letters "E, E, E, E, N, N, N" (this sort of thing seems to happen to us a lot). How many ways are there of arranging these tiles in your rack? In other words, how many seven-letter words (once more in the sense of arbitrary strings of letters) are there that contain exactly three Ns and four Es?

Solution. This is actually pretty simpleminded, but it'll lead to something more interesting in the next problem. The point is, if we think of our seven-letter word as having seven places to fill with Es and Ns, to specify such a word means just to specify which three of the places are to be filled with Ns; or, equivalently, which four are to be filled with Es. The answer is thus

$$\binom{7}{3} = \binom{7}{4} = 35.$$ □

Problem 4.3.5. This problem may not seem like it has much to do with collections, but we'll see in a moment it does. Suppose we live in a city laid out in a rectangular grid, and that our job is located three blocks north and four blocks east of our apartment, as shown in the picture below:

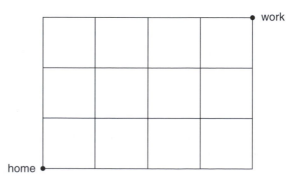

Clearly, we have to walk seven blocks to get to work each morning. But there are many different paths we could take. How many different paths can we take, if we stay on the grid? Think about it before you look at the answer.

Solution. To specify a path from our home to work, we have to give a series of directions like, "Go one block North, then three blocks East, then another block North, then another block East, then another block North," or, for short,

$$N, E, E, E, N, E, N.$$

In other words, paths correspond exactly to words consisting of exactly 3 Ns and 4 Es, and this is exactly the same problem as the last one!

In general, if we have a k-by-l rectangular grid, by the same logic the paths from one corner to the opposite one (with no doubling back) correspond to words formed with k Ns (or Ss) and l Es (or Ws). The number of such paths is thus given by the binomial coefficient

$$\binom{k+l}{k},$$

or, equivalently, the binomial coefficient $\binom{k+l}{l}$. Note the symmetry here: the number of paths in a $k \times l$ grid is the same as the number of paths in an $l \times k$ grid, as the formula verifies.   □

One last item before we conclude this section: we promised in Chapter 2 to show you how to do Problem 2.5.2, and now it's time. We'll start by doing a simpler version of the problem that nonetheless introduces the essential ideas.

Problem 4.3.6. Let's say a class consists of seven orange-shirted students and seven black-shirted students, and for aesthetic reasons we want to line them up so that the colors of the students shirts are alternating. How many ways are there of doing this?

Solution. To start with, the key feature is to separate the problem into two phases:

• First, we have to choose which places in line are to be occupied by orange shirts and which by black shirts. That is, we have to choose a sequence of colors or, if you like, a 14-letter word consisting of seven Os and seven Bs; *but with no two Os adjacent and no two Bs adjacent.* Once we've done that:
• We have to assign an actual student wearing the appropriate shirt color to each place.

In Problem 2.5.1, the first step didn't exist: since there were eight orange shirts and seven black ones, the only possible arrangement of shirt colors was to alternate OBOBOBOBOBOBOBO. Once we realized that, we simply had to assign the eight orange-shirted students to the eight Os in that sequence, and the seven black-shirted students to the seven Bs, for a total of $8! \cdot 7!$ choices.

In the present circumstances, by contrast, there are different possible shirt color sequences. But not many: since we have to alternate colors, the sequence of colors is determined once we specify the first; that is, it's got to be either OBOBOBOBOBOBOB or BOBOBOBOBOBOBO. So there are just 2 choices for the color sequence.

The second step is essentially the same in both Problem 2.5.1 and the present problem. In the present circumstances, once we've decided on a particular arrangement of colors there'll be exactly 7! ways of assigning the seven students wearing orange in class to the seven places in line designated for orange shirts and 7! ways of assigning places to the students wearing black; so there'll be 7! · 7! ways of assigning the 14 students to appropriate places in line. The answer is thus

$$2 \cdot 7! \cdot 7! \ = \ 50{,}803{,}200.$$

Now we're ready to tackle Problem 2.5.2. First, recall the problem: we have a class of six students wearing orange and nine students wearing black, and we want to know how many ways to line them up assuming we don't want any two orange shirts next to each other. If you haven't thought about the problem, take some time now to do so, especially in light of the example we've just worked out.

Ready? Here goes. The first step, just as in the last problem, is to separate the problem into two phases: specifying which places in line are to be occupied by orange shirts and which by black shirts; and then assigning an actual student wearing the appropriate color to each place. Moreover, the second step is essentially the same in both examples: once we've decided on a particular arrangement of colors there'll be exactly 6! ways of assigning the six orange-shirted students in class to the six places in line designated for them, and 9! ways of assigning the black-shirted students' places; so there'll be 6! · 9! ways of assigning the 15 students to appropriate places in line. The answer is thus 6! · 9! times the number of color patterns—that is, 6! · 9! times the number of 15-letter words consisting of 6 Os and 9 Bs, with no two Os in a row.

OK, then, how do we figure out that number? Here is where it gets slightly tricky. The first step is to consider two possibilities: either the sequence ends in a O or it ends in a B. Suppose first that the sequence ends in a B. In that case, we observe, every O in the sequence is necessarily followed by a B. In other words, instead of arranging 6 Os and 9 Bs—subject to the requirement that no O follow another—we can pair off one B with each O to form six OBs, with three Bs left over, and *count arbitrary arrangements of six OBs with three Bs*. We know how many of those there are; it's just

$$\binom{9}{6} = 84.$$

Next, we consider arrangements ending in a O, and we count those similarly: we take the remaining five Os and pair each with a B to form five OBs, with four Bs left over. Again, we can take arbitrary arrangements of these five OBs and four Bs, with a O stuck on at the end; and there are

$$\binom{9}{5} = 126$$

of these. Altogether, then, there are $84 + 126 = 210$ possible color sequences; and since each gives rise to 6! · 9! possible lineups, the total number is

$$210 \cdot 6! \cdot 9! \ = \ 54{,}867{,}456{,}000. \qquad \square$$

Exercise 4.3.7. Let's say that in the small country of Fredonia, the Senate consists of 15 senators, 8 of whom are Republicans and 7 Democrats. By law, therefore, the five-person Ways and Means Committee must consist of three Republicans and two Democrats.

1. How many ways are there of choosing the members of the committee?
2. Suppose that in addition we have to specify a Republican to serve as Chair of the committee and a Democrat to serve as Ranking Member. How many ways are there of choosing the members of the committee, the Chair, and the Ranking Member?
3. Suppose the rules are relaxed and it's required only that the committee not consist entirely of Republicans or entirely of Democrats. How many ways of choosing the members of the committee are there now without specifying a chair or ranking member?

Exercise 4.3.8. Christina's ice cream shop offers only vanilla ice cream, but has 17 different possible toppings to choose from.

1. How many different sundaes can be formed with *exactly* three toppings?
2. How many different sundaes are there with *at least* two toppings?
3. How many different sundaes can be formed with no restriction on the number of toppings?

Exercise 4.3.9. Returning to Problem 4.3.3, suppose now that the 10 players are to be divided randomly into groups of six and four. Is it more likely that Elena and Sofía will be teammates or opponents?

Note: this problem is just slightly trickier than Problem 4.3.3, since in one of the two calculations involved we can't use the multiplication principle. Be sure to check your answer!

Exercise 4.3.10. Suppose an exam has 15 questions: 8 true-false questions and 7 multiple-choice questions. You are asked to answer 5 of the 15 questions, but your professor requires you to answer at least one true-false question and at least one multiple-choice question. How many ways can you choose the questions you plan to answer?

Exercise 4.3.11.

1. Consider the grid below. How many paths of shortest possible length (that is, 13 blocks) are there from the point labeled "home" to the point labeled "work"?

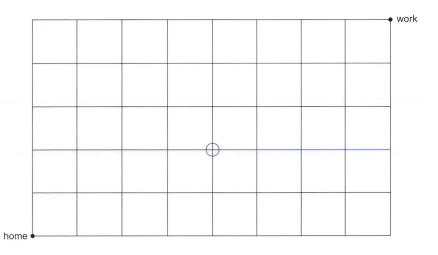

2. Suppose that the small circle on the diagram represents Mike's House of Donuts and Coffee, a crucial stop on the way if you're to arrive at work awake. How many paths from home to work (again of minimal length) pass through Mike's?

3. Suppose by contrast that you're on a strict diet, and that you must at all costs *avoid* passing through the intersection where Mike's is located. Now how many possible paths do you have?

4.4　MULTINOMIALS

Say it's now our job to assign college students to dorm rooms. We have a group of nine students to assign, and three rooms to assign them to: one a quad, one a triple, and one a double. The standard question: how many different ways can we do it?

"That's nothing new," you might say, "we already know how to do it." And you'd be right: to assign the nine students we could start by choosing four of the nine and assign them to the quad. That leaves five students to be assigned to the triple and double; choose three of the remaining five to go to the triple and you're done, since the remaining two necessarily go in the double. Since we had $\binom{9}{4} = 126$ ways of making the first choice and $\binom{5}{3} = 10$ ways of making the second, the answer would be

$$\binom{9}{4} \cdot \binom{5}{3}.$$

But wait: what would happen if we did the assignments in a different order? Suppose, for example, that we started by assigning two of the students to the double, and then chose four of the remaining seven to go the quad? We'd have $\binom{9}{2}$ choices at the first stage, and $\binom{7}{4}$ at the second, so the correct answer would be

$$\binom{9}{2} \cdot \binom{7}{4}.$$

What's up with that?

The answer is, there's no mistake. When we look closer, we see that

$$\binom{9}{4} \cdot \binom{5}{3} = \frac{9!}{4! \, 5!} \cdot \frac{5!}{3! \, 2!}$$
$$= \frac{9!}{4! \, 3! \, 2!},$$

while on the other hand

$$\binom{9}{2} \cdot \binom{7}{4} = \frac{9!}{2! \, 7!} \cdot \frac{7!}{4! \, 3!}$$
$$= \frac{9!}{4! \, 3! \, 2!}.$$

Hence both approaches lead to the same answer, which is equal to 1,260.

So there's really nothing new here. But the sort of numbers that arise here—the number of ways of distributing some number n of objects into three (or more) collections of specified size—are so common that they, like the binomial coefficients, deserve a name and a notation of their own. Think of it this way: maybe the correct (or at any rate the symmetric) way to think of the binomial coefficient $\binom{n}{k}$ is as the number of ways of distributing a group of n objects into two collections, of size k and $n - k$. Well, in the same vein, whenever we have a number n and three numbers a, b, and c that add up to n, we can ask how many ways there are of distributing n objects into three collections, of sizes a, b, and c. We can answer this completely analogously to the way we just did the last problem: we first choose which a of our n objects are to go into the first group; then which b of the remaining $n - a$ are to go into the second. That'll leave c objects, which have to go into the third; so by the multiplication principle the number of ways is

$$\binom{n}{a} \cdot \binom{n-a}{b} = \frac{n!}{a!\,(n-a)!} \cdot \frac{(n-a)!}{b!\,(n-a-b)!}$$

$$= \frac{n!}{a!\,(n-a)!} \cdot \frac{(n-a)!}{b!\,c!}$$

$$= \frac{n!}{a!\,b!\,c!}.$$

This number is called a *multinomial coefficient*, and is typically denoted by the symbol

$$\binom{n}{a,\ b,\ c} = \frac{n!}{a!\,b!\,c!}.$$

Similarly, if a, b, c and d are four numbers adding up to n, the number of ways of distributing n objects into groups of size a, b, c and d is

$$\binom{n}{a,\ b,\ c,\ d} = \frac{n!}{a!\,b!\,c!\,d!},$$

and so on. The most general form of this problem would be: suppose we have n different objects, which we want to distribute into k collections. The number of objects in each collection is specified: the first collection is to have a_1 of the objects, the second a_2 and so on; the kth and last collection is to have a_k of the objects. We ask: how many ways are there of assigning the n objects to the k collections? The answer, as we've suggested, is that:

> The number of ways of distributing n objects into groups of size $a_1, a_2, \ldots, a_k$ is
>
> $$\frac{n!}{a_1! \cdot a_2! \cdot \cdots \cdot a_k!}.$$

Again, the number $\frac{n!}{a_1! \cdot a_2! \cdot \cdots a_k!}$ that appears here is called a multinomial coefficient and denoted $\binom{n}{a_1,a_2,\ldots,a_k}$. Note that in this setting our old friend the binomial coefficient $\binom{n}{k}$ could also be written as $\binom{n}{k,\ n-k}$; but it's easier (and unambiguous) to just drop the $n - k$.

Multinomial coefficients are thus a straightforward generalization of binomial coefficients, and are almost as ubiquitous (though, as we just saw, you don't *really* need to know them: if you just know about binomial coefficients and the multiplication principle, you can solve any problem involving multinomial coefficients).

One classic example of multinomial coefficients is in counting anagrams. By an *anagram* of a word, we mean a rearrangement of its letters: for example, "SAPS" is an anagram of "PASS." (Note that each letter must appear the same number of times in the anagram as in the original word.) In keeping with our conventions, by an anagram we'll mean an arbitrary rearrangement of the letters, not necessarily a word in the English language.

So: how many anagrams does a word have? In some cases this is easy: if a four-letter word, say, has all different letters (that is, none repeated) then an anagram of the word is simply an ordering of its letters, and so there are 4! of them. For example, the word "STOP" has 24 anagrams:

STOP	STPO	SOTP	SOPT	SPTO	SPOT
TSOP	TSPO	TPSO	TPOS	TOSP	TOPS
OSTP	OSPT	OTSP	OTPS	OPST	OPTS
PSTO	PSOT	PTSO	PTOS	POST	POTS

By the same token, an n-letter word with n different letters will have $n!$ anagrams.[1] At the other extreme, the answer is also relatively easy: a word consisting of only one letter repeated n times has no anagrams other than itself; and, as we saw in the example of the Scrabble tiles, if a word consists of k repetitions of one letter and l repetitions of another, it has $\binom{k+l}{k}$ anagrams.

In general, the right way to think about anagrams (from a mathematical point of view) is the way we described in Problem 4.3.4. Suppose, for example, we want to count the anagrams of the word "CHEESES." Any such anagram is again a seven-letter word. If we think of it as having seven slots to fill with the letters C, H, E and S then to specify an anagram we have to specify:

- which one of those seven slots is to be assigned the letter C;
- which one of those seven slots is to be assigned the letter H;
- which three of those seven slots are to be assigned the letter E; and of course:
- which two of those seven slots are to be assigned the letter S.

When we think of it in this way, the answer is clear: it's just the multinomial coefficient

$$\binom{7}{1,\ 1,\ 3,\ 2} = \frac{7!}{1!\,1!\,3!\,2!} = 420.$$

Exercise 4.4.1. How many anagrams does the word "MISSISSIPPI" have? In how many of them are the two Ps next to each other?

Exercise 4.4.2. Consider the following six-letter words: "TOTTER," "TURRET," "RETORT," "PEPPER", and "TSETSE." Which one has the most anagrams, and which the fewest? (You should try and figure out the answer before you actually calculate out the numbers in each case.)

Exercise 4.4.3. How many anagrams does "BOOKKEEPER" have?

---

[1] For the verbally oriented among you, here's a question. Note that, of the 24 rearrangements of the letters STOP, six of them (STOP, SPOT, OPTS, POST, POTS, and TOPS) are actual words in English. Are there four-letter words with more? What five-letter word has the most English word anagrams?

Exercise 4.4.4. It's a beautiful spring day and so you and your 14 siblings have decided to on a picnic. Your uncle who owns a deli has provided sandwiches: 5 ham & cheese, 5 turkey, and 5 egg salad. How many ways are there of distributing the sandwiches to everyone including yourself?

Exercise 4.4.5. The job has fallen to you to assign 18 incoming freshmen to rooms in one particular dormitory. There are six rooms: two quads, two triples, and two doubles.

1. In how many ways can the 18 freshmen be assigned to the rooms?
2. After you submitted your list of assignments from part 1 to the Dean, she complained that some of them put men and women in the same room. If we designate one of the quads, one of the triples, and one of the doubles for women, in how many ways can the rooms be assigned to nine women and nine men?

4.5   SOMETHING'S MISSING

At the Bright Horizons School, prizes are given out to the students to reward excellence. All the prizes are identical, though the school may choose to give more than one prize to a given student.

In Ms. Wickersham's class at Bright Horizons School, there are 14 students: Alicia, Barton, Carolina, and so on up to Mark and Nancy. Ms. Wickersham has eight prizes to award, and has to decide how to give them out—that is, how many prizes each child should get. In how many ways can she do this?

For a slightly different formulation, suppose for the moment that you're the chief distributor for the Widget Transnational Firm. The Widget Transnational Firm has 14 warehouses, called (the Widget Transnational Firm is not a very fanciful outfit) Warehouse A, Warehouse B, and so on up to Warehouse N.

One day, eight containers of widgets show up at the docks, and it's your job to say how many of the eight should go to each of the 14 warehouses. How many ways are there of doing this?

Or: you're in the dining hall one day, and there's a massive fruit bowl, featuring unlimited quantities of each of 14 different fruits: apples, bananas, cherries, and so on up to nectarines. Feeling a mite peckish, you decide to help yourself to eight servings of fruit, possibly taking more than one serving of a given kind. How many different assortments can you select?

Well, what is the point here? Actually, there are a couple: one, we don't know how to solve this problem, and two, we should. Think about it: we've derived, so far in this book, three formulas for counting the number of ways of making a series of k selections from a pool of n objects:

- We know the number of ways of choosing a sequence (that is, the order does matter) from the pool, with repetitions allowed: it's n^k.
- We know the number of ways of choosing a sequence from the pool, with no repetitions: it's $\frac{n!}{(n-k)!}$.
- We know the number of ways of choosing a collection (that is, the order doesn't matter) from the pool, with no repetitions: it's $\binom{n}{k} = \frac{n!}{k!(n-k)!}$.

If we arrange these formulas in a table, as here:

	repetitions allowed	without repetitions
sequences	n^k	$\dfrac{n!}{(n-k)!}$
collections	??	$\dfrac{n!}{k!\,(n-k)!}$

it's clear that something's missing: we don't have a formula for the number of collections of k objects, chosen from a pool of n, with repetitions allowed. That's what all those problems we just listed (or that single problem we repeated three times) involve.

So: are we going to tell you the answer, already? Well, yes and no. We are going to work out the formula in Chapter 7, at the end of this part of the book. But we thought it'd be nice to leave you something to think about and work on in the meantime. So we'll leave it as a challenge: can you solve the problem(s) above before we get to Chapter 7?

5 Games of chance

A large part of the reason why we are so interested in counting collections is that is will allow us to compute the probabilities of certain events occurring, at least in contexts where the different possible outcomes are equally likely. In this chapter we're going to use our counting skills to discuss some basic problems in probability. We'll focus primarily on games—flipping coins, rolling dice, and playing poker and bridge—but it should be clear how the same ideas can be applied in other areas as well.

5.1 FLIPPING COINS

Suppose we flip a coin six times. What's the probability of getting three heads and three tails?

To answer this, we have to start with two hypotheses. The first is simply that we have a fair coin—that is, one that on average will come up heads half the time and tails half the time.

To express the second hypothesis, we have to introduce one bit of terminology. By the *outcome* of the process of flipping the coin six times we mean the sequence of six results, which we can think of as a six-letter word consisting of Hs and Ts. How many possible outcomes are there? That's easy: by the very first formula we worked out using the multiplication principle, the number of such sequences is 2^6, or 64.

Now, it is a basic assumption of probability that *all 64 outcomes are equally likely.* In effect, this means simply that the result of each coin flip is equally likely to be heads or tails, irrespective of what the result of the previous flips might have been—a slogan is, "the coin has no memory." We should emphasize here that this is really a hypothesis: even though you might think of this as self-evident, and it's been verified extensively by experiment, it's not something we can logically prove. Indeed, there were long periods of human history when just the opposite was thought to be true: when people (and not just degenerate gamblers) believed, for example, that after a long run of heads a tail was more likely than another head. In Part III, when we study probability distributions associated to repetitions of the same game, we'll get a sense of why many people persist in this mistaken belief.

So: let's adopt these hypotheses:

- all coins are fair (unless otherwise specified);
- the result of each coin flip is independent of the results of each other coin flip.

What they mean is that any specific outcome—three heads followed by three tails (HHHTTT), or three tails followed by three heads (TTTHHH), and so on—will occur $1/64^{\text{th}}$ of the time; in other words, the probability is 1 in 64 that any specified outcome will occur on a given experiment of six flips. Given this, to determine the likelihood of getting exactly three heads and three tails on a given six flips, we have to answer a counting problem: of the 64 possible outcomes, how many include three Hs and three Ts?

This is also an easy problem: the number of six-letter words consisting of three Hs and three Ts is just the binomial coefficient

$$\binom{6}{3} = 20.$$

Now, if each of these outcomes occurs $1/64^{\text{th}}$ of the time, then in the aggregate we would expect one of these 20 outcomes to occur

$$\frac{20}{64} = \frac{5}{16} = .3125$$

of the time. In other words, when we flip six coins, we expect to get an equal number of heads and tails a little less than one-third of the time. This rather counterintuitive finding is what motivated your friend's bar bet, where you had to buy the next round if you couldn't predict *exactly* how many heads would come up in six coin flips.

You can probably see the general rule here: if we flip a coin n times, there are 2^n possible outcomes—corresponding to the n-letter words consisting entirely of Hs and Ts—each of which will on average occur 1 in 2^n times. The number of these outcomes that involve exactly k heads and $n - k$ tails is $\binom{n}{k}$; and so our conclusion is that:

> The probability of getting exactly k heads in n flips is $\dfrac{\binom{n}{k}}{2^n}$.

We'll look further into other probabilities associated with flipping coins, but before we do we should take a moment to point out that this is the basic paradigm of probability: in general, when all possible outcomes of an experiment or process are equally likely, and we separate out the collection of outcomes into two kinds—the so-called "favorable" and "unfavorable"—the probability of a favorable outcome is simply

$$\text{probability of a favorable outcome} = \frac{\text{number of favorable outcomes}}{\text{total number of possible outcomes}}.$$

Note again that this presupposes that all outcomes are equally likely. If that's not the case, or if we define "outcome" incorrectly, all bets are off, so to speak.

Here are some more examples. As with all probability problems, it's fun to think about them a little and try to estimate the probabilities before you actually go ahead and calculate them: sometimes they can surprise you (and you can come up with some lucrative bets).

Problem 5.1.1. Let's say you flip a coin eight times. What is the probability of getting three or more heads?

Solution. We have to figure out, of the $2^8 = 256$ eight-letter words consisting entirely of Hs and Ts, how many have at least three Hs. It's slightly easier to figure out how many don't, and use the subtraction principle: we have to count the number of such words that have zero, one or two Hs, and by what we've done the number is

$$\binom{8}{0} + \binom{8}{1} + \binom{8}{2} = 1 + 8 + 28 = 37.$$

The number of such sequences that do have three or more heads is thus $256 - 37 = 219$, and so the probability of getting at least three heads is

$$\frac{219}{256} \sim .85.$$

In other words, you'll get three or more heads in eight flips about 85% of the time. □

Problem 5.1.2. Say you and a friend are gambling. You flip nine coins; if they split 4/5—that is, if they come up either four heads and five tails or four tails and five heads—you pay him \$1; otherwise, he pays you \$1. Who does the game favor?

Solution. There are $2^9 = 512$ possible outcomes of the nine coin flips, of which

$$\binom{9}{4} + \binom{9}{5} = 126 + 126 = 252$$

involve either four or five heads. That leaves

$$512 - 252 = 260$$

outcomes that don't. Thus the odds are (very slightly) in your favor. □

Problem 5.1.3. A variant of the last problem. You and your friend flip six coins; if three or more come up heads you pay him \$1; if two or fewer are heads, he pays you \$1. Who does the game favor?

Solution. We start the same way as the last problem: figuring out, of the $2^6 = 64$ possible outcomes of the six coin flips, how many result in a win for you and how many in a win for your friend. First, the number of outcomes with fewer than three heads—that is, with 0, 1 or 2 heads—is the sum

$$\binom{6}{0} + \binom{6}{1} + \binom{6}{2} = 1 + 6 + 15 = 22.$$

On these 22 outcomes, you win \$1. That leaves

$$64 - 22 = 42$$

outcomes where your friend wins \$1. You'll lose \$1, in other words, slightly less than twice as often as you win \$1, so the odds are clearly in your friend's favor. □

The game considered in Problem 5.1.3 would be much more fair if the payoffs were changed so that your friend paid you \$2 whenever two or fewer coins come up heads, while you still paid only $1 if three or more come up heads. With these payoffs, you'd be slightly better off, since you'd lose $1 slightly less than twice as often as you'd win $2.

In Chapter 8 we'll learn how to compute the *expected value* of games with different payouts occurring with varying probabilities. Even though we haven't formally introduced this notion, now is as good a time as any to attempt the following exercise.

Exercise 5.1.4. You and a friend play the following game. You each flip three coins, and whoever gets more heads wins; if you get the same number, you win. If you win, your friend pays you $1; if your friend wins, you pay her $2. Who does the game favor?

Exercise 5.1.5.

1. Say you flip a fair coin five times. What's the probability that some three flips in a row will come up the same?
2. Say you flip a fair coin eight times. What is the probability of getting 4 heads and 4 tails in any order?
3. Say you flip a fair coin ten times. What is the probability of getting between 4 and 6 heads?
4. Say you flip a fair coin seven times. What is the probability you'll get either 3 heads and 4 tails, or 4 heads and 3 tails?

5.2 ROLLING DICE

As far as mathematics goes, dice are not that different from coins; this section won't introduce any new ideas. But because dice have six faces rather than two, and you can do things like add the results of several dice rolls, they're more interesting. (Las Vegas casinos have tables for playing craps, a game of rolling dice, but they don't have coin-flipping tables.)

Let's start by rolling two dice and calculating some simple probabilities. Again, the hypotheses: first, the dice are fair; in other words, each of the six faces will on average come up one-sixth of the time. Second, if we define the outcome of the roll to be a sequence of two numbers between 1 and 6, the $6 \times 6 = 36$ possible outcomes are all equally likely, that is, each occurs $\frac{1}{36}$ of the time.

We should stop for a moment here and try to clarify one potential misunderstanding. Most of the time, when we roll a pair of dice, the two dice are indistinguishable and we don't think of one as "the first die" and the other as "the second die." *But for the purposes of calculating probabilities, we should.* For example, there are two ways of rolling "a 3 and a 4": the first die could come up 3 and the second 4, or vice versa; "a 3 and a 4" thus comes up 2/36, or 1/18, of the time. By contrast, "two 3s" arises in only one way, and so occurs only 1/36 of the time. This can be confusing, and even counterintuitive: when we roll two identical dice, we may not even know whether we've rolled "a 3 and a 4" or "a 4 and a 3." It may help to think of the dice as having different colors—one red and one blue, say—or of rolling them one at a time, rather than together.

With this said, let's calculate some probabilities. To begin with, let's say we roll two dice and add the numbers showing. We could ask, for example: what's the probability of rolling a 7?

To answer that, we simply have to figure out, of the 36 possible outcomes of the roll, how many yield a sum of 7? This we can figure out by hand: we could get 1 and 6, 2 and

5, 3 and 4, 4 and 3, 5 and 2, or 6 and 1, for a total of six outcomes. The probability of rolling a 7, accordingly, is 6/36, or 1/6.

By contrast, there is only one way of rolling a 2—both dice have to come up 1—so that'll come up only 1/36 of the time. Similarly, there are two ways of rolling a 3—a 1 and a 2, or a 2 and a 1—so that arises 2/36, or 1/18 of the time. You can likewise figure out of the probability of any roll; you should take a moment and verify the probabilities in the table below.

sum	number of ways to achieve the sum	probability
2	1	1 in 36
3	2	1 in 18
4	3	1 in 12
5	4	1 in 9
6	5	5 in 36
7	6	1 in 6
8	5	5 in 36
9	4	1 in 9
10	3	1 in 12
11	2	1 in 18
12	1	1 in 36

Now let's look at some examples involving three or more dice.

Problem 5.2.1. Suppose now you roll three dice. What is the probability that the sum of the faces showing will be 10? What is the probability of rolling a 12? Which is more likely?

Solution. There are 6^3, or 216, possible outcomes of the roll of three dice; we just have to figure out how many add up to 10, and how many add up to 12.

There are many ways of approaching this problem—we could even just write out all the possible outcomes, but it's probably better to be systematic. Here's one way: if the sum of the first two rolls is any number between 4 and 9, then there is one and only one roll of the third die that will make the sum of all three equal to 10. Thus, the number of ways we can get 10 is simply the sum of the number of outcomes of two dice rolls that add up to 4, the number of outcomes of two dice rolls that add up to 5, and so on up to the number of outcomes of two dice rolls that add up to 9. We worked all these out a moment ago; the answer is

$$3 + 4 + 5 + 6 + 5 + 4 = 27.$$

Thus the probability of rolling a 10 with three dice is 27 out of 216, or simply 1 in 8.

Similarly, the number of ways we can get 12 is simply the sum of the number of outcomes of two dice rolls that add up to 6, the number of outcomes of two dice rolls that add up to 7, and so on up to the number of outcomes of two dice rolls that add up to 11; that is,

$$5 + 6 + 5 + 4 + 3 + 2 = 25.$$

So the probability of rolling a 12 with three dice is slightly less than the probability of rolling a 10. □

Problem 5.2.2. Let's again roll three dice; this time, calculate the probability of getting at least one 6.

Solution. This is actually simpler than the last problem, because it's easier to be systematic. Just use the subtraction principle: the number of outcomes that include at least one 6 is 216 minus the number of outcomes that don't involve a 6; that is,

$$216 - 5^3 = 216 - 125 = 91.$$

The probability of getting at least one 6 on three rolls is thus 91 out of 216. □

Problem 5.2.3. For a final example, let's roll seven dice. What is the probability of getting exactly two 6s?

Solution. This time, it's the multiplication principle we want to use. We know there are $6^7 = 279,936$ sequences of seven numbers from 1 to 6; we have to count how many such sequences contain exactly two 6s. Well, we can specify such a sequence by choosing, in turn:

- which two of the seven numbers in the sequence are to be the 6s, and
- what the other five numbers in the sequence are to be.

For the first, the number of choices is just $\binom{7}{2}$, or 21. The second involves simply specifying a sequence of five numbers other than 6, that is, from 1 to 5; the number of choices is thus $5^5 = 3,125$. The total number of the sequences we're counting is thus

$$21 \times 3,125 = 65,625$$

and the probability of rolling such a sequence is

$$\frac{65,625}{279,936} \sim .234.$$

In other words, our chances are slightly less than 1 in 4. □

Exercise 5.2.4. Say you roll five dice.

1. What is the chance that you'll get at least one 5 and at least one 6?
2. What is the probability that the *sum* of the numbers showing is 5 or less?

Exercise 5.2.5. In the game of Phigh, each player rolls three dice; their score is the highest number that appears.

1. What is the probability of scoring 1?
2. What is the probability of scoring 2?
3. What is the probability of scoring 2 or less?
4. Your opponent scored 4. What is the probability that you'll win (that is, score 5 or 6)?

Exercise 5.2.6. Suppose you roll a fair die 10 times.

1. What's the probability that you'll roll a "7" exactly twice?

2. Which of the following is most likely:
    (a) you roll exactly one 6;
    (b) you roll exactly two 6s;
    (c) you roll exactly three 6s;
    (d) you roll no 6s.

PLAYING POKER

It's time to graduate from dice to cards, and we're going to focus here primarily on probabilities associated with poker.

To start with, let's establish the rules. A standard deck consists of 52 cards. There are four suits: spades (♠), hearts (♡), diamonds (♢), and clubs (♣). There are 13 cards of each suit, with denominations 2, 3, 4, 5, 6, 7, 8, 9, 10, jack (J), queen (Q), king (K), and ace (A). A poker hand consists of five cards and the game is won by the player whose hand has the highest "rank." The ranks of the various hands are listed below in order from lowest (most likely) to highest (least likely):

- A pair: a hand including two cards of the same denomination.
- Two pair: a hand including two cards each of two denominations.
- Three of a kind: a hand including three cards of the same denomination.
- Straight: a hand in which the denominations of the five cards form an unbroken sequence. For this purpose an ace may be either high or low; that is, A 2 3 4 5 and 10 J Q K A are both straights.
- Flush: a hand in which all five cards belong to the same suit.
- Full house: a hand consisting of three cards of one denomination and two cards of another denomination.
- Four of a kind: a hand including four cards of the same denomination.
- Straight flush: a hand consisting of five cards of the same suit forming an unbroken sequence.

Note that when we talk about a hand whose rank is "exactly three of a kind," we'll mean a hand of that rank and no higher.

We're going to start with the basic question: if you're dealt five cards at random, what are the chances of getting a given type of hand or better? Here "at random" means that all the possible hands, of which there are

$$\binom{52}{5} = \frac{52 \cdot 51 \cdot 50 \cdot 49 \cdot 48}{5 \cdot 4 \cdot 3 \cdot 2 \cdot 1} = 2{,}598{,}960,$$

are equally likely to arise; so that the probability of getting a particular type of hand are just the total number of such hands divided by 2,598,960. Our goal, then, will be (as usual) to count the number of hands of each type.

We'll start at the top, with straight flushes. These are straightforward to count via the multiplication principle: to specify a particular straight flush, we simply have to specify the denominations and the suit, which are independent choices. There are four suits, obviously; and as for the denominations, a straight can have as its low card any card from A up to 10 (remember that we count A 2 3 4 5 as a straight), so there are 10 possible denominations. There are thus

$$4 \times 10 = 40$$

straight flushes, and the probability of being dealt one in five cards is accordingly

$$\frac{40}{2,598,960} \sim .0000153,$$

or approximately 1 in 64,974. Not an everyday occurrence: if you play, for example, on the order of two hundred hands a week, it'll happen to you roughly once in six or seven years.

Next is four of a kind. Again, the multiplication principle applies more or less directly: to specify a hand with four of a kind, we have to specify first the denomination of the four, and then say which of the remaining 48 cards in the deck will be the fifth card of the hand. The number of choices is accordingly

$$13 \times 48 = 624,$$

and the probability of being dealt one in five cards is accordingly

$$\frac{624}{2,598,960} \sim .00024001$$

or, in cruder terms, approximately 1 in 4,000. A good bit more likely than a straight flush, in other words, but don't hold your breath; again, if you play on the order of two hundred hands a week, on average you'll get two or three of these a year.

Note that if we wanted to calculate the probability of getting "four of a kind or better" we'd have to add the number of hands with four of a kind and the number of hands with a straight flush. In general, we're going to calculate here the likelihood of getting a hand of exactly a given rank; to count the number of hands of a specified rank *or higher* you'll have to add up the numbers of hands of each rank above.

Full houses are also straightforward to count. Since a full house consists of three cards of one denomination and two cards of another, we have to specify first the denomination of which we have three cards, and the denomination of which we have two; then we have to specify which three of the four cards of the first denomination are in the hand, and which two of the second. Altogether, then, the number of choices is

$$13 \times 12 \times \binom{4}{3} \times \binom{4}{2} = 13 \times 12 \times 4 \times 6$$
$$= 3,744.$$

The probability is

$$\frac{3,744}{2,598,960} \sim .0014406,$$

or approximately 1 in 700.

Flushes are even easier to count: we specify a suit (4 choices), and then which five of the 13 cards in that suit will constitute the hand. The number of flushes is thus

$$4 \times \binom{13}{5} = 4 \times 1,287$$
$$= 5,148.$$

Remember, though, that this includes straight flushes too! If we want to count the number of hands with *exactly* a flush and not any higher rank, we have to subtract those 40 hands, so that the number is

$$5{,}148 - 40 = 5{,}108.$$

The probability is thus

$$\frac{5{,}108}{2{,}598{,}960} \sim .0019654,$$

or approximately 1 in 500.

If you're with us so far, straights are likewise simple to count: we have to specify the denominations of the cards—10 choices, as we counted a moment ago—and then the suits, which involve four choices for each of the five cards. The total number of straights is therefore

$$10 \times 4^5 = 10 \times 1{,}024$$
$$= 10{,}240;$$

and if we exclude straight flushes, the number of hands whose rank is exactly a straight is

$$10{,}240 - 40 = 10{,}200$$

The probability is

$$\frac{10{,}200}{2{,}598{,}960} \sim .0039246,$$

or very approximately 1 in 250.

Next, we count hands with exactly three of a kind. Initially, this is similar to the cases we've done before: we have to specify the denomination of the three (13 choices); which three of the four cards of that denomination are to be in the hand ($\binom{4}{3} = 4$ choices), and finally which two of the remaining 48 cards of the deck will round out the hand.

But here there's one additional wrinkle: since we're counting only hands with the rank "three of a kind," and not "full house," the last two cards can't be of the same denomination. Now, if we were counting the number of sequences of two cards of different denominations from among those 48, the answer would be immediate: we have 48 choices for the first, and 44 for the second, for a total of $48 \times 44 = 2{,}112$ choices. Since the order doesn't matter, though, and because each collection of two such cards corresponds to two different sequences, the number of pairs of cards of different denominations from among those 48 is

$$\frac{48 \cdot 44}{2} = 1{,}056.$$

The number of hands with exactly three of a kind is thus

$$13 \times 4 \times 1{,}056 = 54{,}912$$

and the probability is

$$\frac{54{,}912}{2{,}598{,}960} \sim .021128$$

or roughly 1 in 50. In other words, if your typical night of poker consists of a couple hundred hands, you're likely to be dealt three of a kind four times.

Counting hands with exactly two pair is slightly easier. We specify the two denominations involved; we have $\binom{13}{2} = 78$ choices there. Then we have to say which two of the four cards of each of these denominations go in the hand; that's

$$\binom{4}{2}^2 = 6^2 = 36$$

choices. Finally, we have to say which of the remaining 44 cards in the deck (of the other 11 denominations) will complete the hand. The total number is, accordingly,

$$78 \times 36 \times 44 = 123{,}522$$

and the probability is

$$\frac{123{,}522}{2{,}598{,}960} \sim .047539,$$

or approximately 1 in 20.

Finally, we come to the hands with exactly one pair. We can do this in a similar fashion to our count of hands with three of a kind; choose the denomination of the pair (13); choose two cards of that denomination ($\binom{4}{2} = 6$), and finally choose three cards among the 48 cards not of that denomination. But, as in the case of hands with three of a kind, there's a wrinkle in that last step: the three cards not part of the pair must all be of different denominations. Again, if we were counting sequences, rather than collections, of cards, this would be straightforward: there'd be $48 \times 44 \times 40$ choices. Since we have to count collections, however, and since every collection of three cards corresponds to $3! = 6$ different sequences, the number of such collections is

$$\frac{48 \cdot 44 \cdot 40}{6} = 14{,}080.$$

The number of hands with exactly a pair is thus

$$13 \times 6 \times 14{,}080 = 1{,}098{,}240$$

and the probability is

$$\frac{1{,}098{,}240}{2{,}598{,}960} \sim .42256,$$

or a little worse than half. As we remarked before, if we want to find the probability of being dealt a pair *or better*, we have to add up the numbers of all hands better than a pair: the total is

$$36 + 624 + 3{,}744 + 5{,}112 + 10{,}200 + 54{,}912 + 123{,}552 + 1{,}098{,}240 = 1{,}296{,}420.$$

Now, there's another way to calculate this number, and it gives us a way to check a lot of our calculations. We can count the number of hands with a pair or better by the subtraction principle: that is, count the hands with no pairs, and subtract that from the total number of hands. To count the hands with no two cards of the same denomination, as in the calculation we did of hands with three of a kind or with a pair, we count first the sequences of five cards, no two of the same denomination; this number is simply

$$52 \times 48 \times 44 \times 40 \times 36.$$

But every such hand of cards corresponds to $5! = 120$ sequences, so the number of hands with no two cards of the same denomination is

$$\frac{52 \times 48 \times 44 \times 40 \times 36}{120} = 1{,}317{,}888.$$

But we're not quite done: these 1,317,888 hands include straights, flushes and straight flushes, and if we want to count hands that *rank below* a pair, we have also to exclude these. Thus the total number of poker hands ranking below a pair will be

$$1{,}317{,}888 - 40 - 5{,}108 - 10{,}200 = 1{,}302{,}540,$$

and the number of hands ranked a pair or better will be

$$2{,}598{,}960 - 1{,}302{,}540 = 1{,}296{,}420,$$

as we predicted. Note that the probability of getting a pair or better is thus

$$\frac{1{,}296{,}420}{2{,}598{,}960} \sim .49843,$$

or very nearly one in two.

Exercise 5.3.1. What is the probability of being dealt a *busted flush*—that is, four cards of one suit and a fifth card of a different suit?

Exercise 5.3.2. What are the probabilities that a five-card poker hand will contain at least one ace?

Exercise 5.3.3. In a standard deck of cards, two suits (diamonds and hearts) are red, and the other two (spades and clubs) are black. We'll call a poker hand a *nearsighted flush* if all five cards are of the same color.

1. What is the probability of being dealt a nearsighted flush?
2. Is this the same as the probability that 5 coins flipped will all come up the same? Why, or why not?

5.4 REALLY PLAYING POKER

This section is probably unnecessary, but our lawyers insisted that we include it.

The probabilities we've just calculated are obviously relevant to playing poker, but they're only the tip of the tip of the iceberg. In almost all versions of poker, your hand isn't simply dealt to you all at once; it comes in stages, after each of which there's a round of betting. Each time you have to bet, you have to calculate the likelihood of winding up with each possible hand, based on what you have already and how many cards you have yet to receive.

What's more, most poker games involve at least some cards dealt face up, and every time a card is dealt face up, it changes the odds of what you're likely to receive on succeeding rounds, and what your hand is likely to wind up being. In addition, every time someone bets (or doesn't) or raises (or doesn't), it changes the (estimated) odds of what their hole cards are, and hence what cards you're apt to be dealt on succeeding rounds. In fact, every time it's your bet, you have to calculate the probability of your achieving each possible hand, and the amount you stand to win or lose depending on what you get (which depends in turn on other factors: what the other players get, how much is currently in the pot, how much the other players will contribute to the pot, and how much you'll have to contribute to the pot).

To be really good at poker, you have to be able to calculate these probabilities accurately (and unsentimentally). At the same time, it can never be exact: for one thing, no one can make that many calculations that quickly. For another, figuring out how likely it is that the player across the table really does have a king under is necessarily an inexact science. In other words, serious poker exists somewhere in that gray area between mathematics and intuition. Those of us with weaknesses in either field should probably limit our bets.

5.5   BRIDGE

Bridge is another card game that calls for estimations of probabilities. We're not going to discuss the game in any depth or detail at all, but there is one aspect of the game that makes for a beautiful problem in probability, which we'll describe.

In bridge, each player is dealt a hand of 13 cards from a standard deck. This means there are

$$\binom{52}{13} = 635{,}013{,}559{,}600$$

possible hands, all of which we're going to assume are equally likely on any given deal.[1] Now, every hand has what is called a *distribution*, meaning how many cards it has from the four different suits: for example, a hand with four cards of one suit and three each of the others is said to have a 4333 distribution; a hand with four cards each of two suits, three of a third, and two of the final suit is said to have a 4432 distribution; and so on.

The question we want to take up here is: what is the probability that a bridge hand has a given distribution? As a special case, we could ask: which is more likely to occur, a 4333 distribution, or a 4432? How do the probabilities of either compare to the likelihood of getting a relatively unbalanced distribution, like 5431?

Let's start by counting the number of hands with a 4333 distribution. Basically, we can specify such a hand in two stages: first, we specify which suit is to be the four-card suit, and then we have to specify which of the 13 cards from each suit we're to receive. To specify the four-card suit, there are clearly 4 choices; and as for specifying which of the 13 cards from each suit we're to receive, we have to choose four cards from one suit and three from each of the others. By the multiplication principle, then, the total number of choices is

---

[1]   This is not actually the case at many bridge tournaments, where the tournament administrator artificially deals rare and interesting hands to be played at each table, with the competitors rotating between tables.

$$4 \cdot \binom{13}{4} \cdot \binom{13}{3} \cdot \binom{13}{3} \cdot \binom{13}{3}$$

or, in factorials,

$$= 4 \cdot \frac{13!}{4!\,9!} \cdot \frac{13!}{3!\,10!} \cdot \frac{13!}{3!\,10!} \cdot \frac{13!}{3!\,10!},$$

which works out to

$$= 66{,}905{,}856{,}160.$$

The probability of being dealt a hand with a 4333 distribution is thus

$$\frac{66{,}905{,}856{,}160}{635{,}013{,}559{,}600} \sim .105,$$

or slightly better than one in ten.

Let's do the 4432 distribution next. The idea is the same: first we figure out how many ways we can match the four numbers with the four suits; then, once we've specified how many cards of each suit we're to receive, we calculate how many ways we can choose those cards. For the first part, we have to choose the suit with two cards (four choices) and then the suit with three (three choices); the remaining two suits will each get four cards. The total number of choices is thus

$$4 \cdot 3 \cdot \binom{13}{4} \cdot \binom{13}{4} \cdot \binom{13}{3} \cdot \binom{13}{2}$$
$$= 4 \cdot 3 \cdot \frac{13!}{4!\,9!} \cdot \frac{13!}{4!\,9!} \cdot \frac{13!}{3!\,10!} \cdot \frac{13!}{2!\,11!}$$
$$= 136{,}852{,}887{,}600.$$

The probability of being dealt a hand with a 4432 distribution is thus

$$\frac{136{,}852{,}887{,}600}{635{,}013{,}559{,}600} \sim .216,$$

or slightly better than one in five. So in fact we see that you're more than twice as likely to be dealt a hand with a 4432 distributions as one with a 4333 distribution!

By now you've probably got the idea, so you can do some yourself:

Exercise 5.5.1. Guess which of the following distributions is most likely and then calculate the probabilities:

1. a 5332 distribution;
2. a 4441 distribution; and
3. a 7321 distribution.

Exercise 5.5.2. What is the probability that a random bridge hand has exactly 7 cards of one suit?

The next exercise has to do with a basic problem in bridge: once you've seen your cards, what are the factors that govern what everyone else's hand looks like? Clearly what you've got has some effect on the probabilities: if you have 11 spades, for example, you can be comfortably certain that no one at the table has a 4333 distribution. It's a hard problem, but if you can do it you can call yourself a master counter.

Exercise 5.5.3. Say you're playing bridge, and you pick up your hand to discover you have a 7321 distribution. What is the chance that the player to your left has a 4333 distribution?

5.6 **THE BIRTHDAY PROBLEM**

Everyone has a birthday; and, leaving aside for the moment those unfortunate souls born on February 29 of a leap year, everyone's birthday is one of the 365 days of the standard year. The probability of two people selected at random having the same birthday is, accordingly, 1 in 365.

So, suppose now we get 10 people together at random. What is the probability that two have the same birthday? How about a group of 25, or 50, or 100? It's pretty clear that as the number of people in the group increases, so does the probability of two people having the same birthday—when you get up to 366 people, of course, it's a lock—so we might ask: *for what size group is there actually a better than 50% chance of two people having the same birthday*? We know how to calculate the probabilities by now, of course, but before we do so you might want to take a few moments out and think about it—take a guess.

Time's up; here we go. Suppose we line up a group of, say, 50 people, and list their birthdays. We get a sequence of 50 days of the year; and assuming the people were picked randomly—so that each one is as likely to have been born on one day as another—of the 365^{50} possible such sequences, all are equally likely.

So: how many of these 365^{50} possible sequences involve a repeated day? Well, we know how many don't: the number of sequences of 50 days without repetition is, by the standard formula, the product

$$\frac{(365)!}{(315)!} \quad \text{or} \quad 365 \cdot 364 \cdot 363 \cdots 317 \cdot 316.$$

The probability of there *not* being a repeated birthday among 50 people is thus

$$\frac{365 \cdot 364 \cdot 363 \cdots 317 \cdot 316}{365^{50}}.$$

Now, these are some hefty numbers, and we have to be careful how we multiply them out: if we just ask our calculator to come up with 365^{50}, it'll never speak to us again. But we can rewrite this in a form that keeps the numbers reasonably sized:

$$\frac{365 \cdot 364 \cdot 363 \cdots 317 \cdot 316}{365^{50}} = \frac{365}{365} \cdot \frac{364}{365} \cdots \frac{317}{365} \cdot \frac{316}{365}$$

$$= 1 \cdot \left(1 - \frac{1}{365}\right) \cdot \left(1 - \frac{2}{365}\right) \cdots \left(1 - \frac{48}{365}\right) \cdot \left(1 - \frac{49}{365}\right).$$

This is something we (or rather our computers) can evaluate, and the answer is that the probability of not having a repeat is 0.0296. In other words, if we take 50 people at random, the probability is better than 97% that two will share a birthday! Pretty surprising, when you think about it.

In fact, if you work it out, it's already the case with 23 people that the probability of a repeated birthday is 50.7%, or better than half; and by the time you get to 30 people the probability is 70.6% that two people will have the same birthday.

Exercise 5.6.1. How many people need to be in a room in order for it to be more likely than not that two of them share the same zodiac sign (of which there are 12)?

Exercise 5.6.2. How many people need to be in a room (not including yourself) in order for it to be more likely than not that one of them shares your birthday?

Interlude

If you're with us so far—if most of the calculations in the last chapter make sense to you—you've got a pretty good idea of what counting is about. In particular, you've seen all the ideas and techniques of counting that we're going to use in the rest of this book. From a strictly logical point of view, you could proceed directly to Part II to continue the study of probability commenced in Chapter 5.

But in the course of our counting, we've come across a class of numbers, the binomial coefficients, that are worth studying in their own right, both for the fascinating properties and patterns they possess and for the way they crop up in so many areas of mathematics. We're going to take some time out here, accordingly, and devote a chapter to the binomial coefficients themselves. We'll follow this up in Chapter 7 by discussing some more advanced counting techniques, before returning to the study of probability theory in Part II.

These detours are common in mathematics—the tools that we develop to solve a particular problem often open up surprising areas of investigation in their own right.

6 Pascal's triangle and the binomial theorem

Probably the best way to go about looking for patterns in binomial coefficients is simply to make a table of them and stare at it—maybe we'll be able to deduce something. (Mathematicians like to give the impression that they arrive at their conclusions by abstract thought, but the reality is more prosaic: most of us at least start with experimentation.) As for the form this table should take, there's a classic way of representing the binomial coefficients that is particularly well-suited to displaying their patterns, called *Pascal's triangle*.

Pascal's triangle consists of a sequence of rows, where each row gives the values of the binomial coefficients $\binom{n}{k}$ as k increases for a particular value of n. For example, the row with $n = 1$ has only two numbers in it:

$$\binom{1}{0} = 1 \quad \text{and} \quad \binom{1}{1} = 1.$$

The row with $n = 2$ has three:

$$\binom{2}{0} = 1, \quad \binom{2}{1} = 2, \quad \binom{2}{2} = 1.$$

The $n = 3$ row has four:

$$\binom{3}{0} = 1, \quad \binom{3}{1} = 3, \quad \binom{3}{2} = 3, \quad \binom{3}{3} = 1,$$

and so on. These rows are arranged one above the other, with the centers vertically aligned. The whole thing looks like:

Of course, the triangle continues forever, but we have to stop somewhere. Note that we start with the row for $n = 0$, consisting of the one binomial coefficient $\binom{0}{0} = 1$. (In general, we'll refer to the row starting with $\binom{n}{0}$, $\binom{n}{1}$, etc., as the nth row.)

As we said, the patterns we've observed so far in the binomial coefficients are all in evidence here. The most striking is the symmetry: if we flip the whole thing around its central vertical axis, the triangle is unchanged. This is, naturally, a reflection of the fact that

$$\binom{n}{k} = \binom{n}{n-k}.$$

Table 6.1: Pascal's triangle.

					1						$(n = 0)$
				1		1					$(n = 1)$
			1		2		1				$(n = 2)$
		1		3		3		1			$(n = 3)$
	1		4		6		4		1		$(n = 4)$
1		5		10		10		5		1	$(n = 5)$
1	6		15		20		15		6	1	$(n = 6)$
1	7	21		35		35		21	7	1	$(n = 7)$
1	8	28	56		70		56	28	8	1	$(n = 8)$

We also see that

$$\binom{n}{0} = \binom{n}{n} = 1$$

for any n in the fact that the edges of the triangle consist entirely of 1s, and that

$$\binom{n}{1} = \binom{n}{n-1} = n$$

in the fact that the second (and second-to-last) entry in each row is the row number.

All the laws we've discussed so far have to do with the entries on a single row: the first and last are 1, the row is symmetric, and so on. But writing out all the binomial coefficients together reveals a pattern among entries in different rows as well. It's obvious once you think of it, and maybe you've seen it before; if you haven't, take a moment out to stare some more at the triangle before we point it out.

OK, here it is: *Each entry in the table is exactly the sum of the two entries closest to it in the row immediately above it.* The entry $\binom{5}{2} = 10$ in the $n = 5$ row is the sum of the two entries $\binom{4}{1} = 4$ and $\binom{4}{2} = 6$ in the $n = 4$ row; the entry $\binom{8}{3} = 56$ in the $n = 8$ row is the sum of the two entries $\binom{7}{2} = 21$ and $\binom{7}{3} = 35$ in the $n = 7$ row, and so on. In fact, you can use this pattern to write down the next ($n = 9$) row of the triangle without actually multiplying and dividing any more factorials, but just adding pairs of terms in the $n = 8$ row:

						1						
					1		1					
				1		2		1				
			1		3		3		1			
		1		4		6		4		1		
	1		5		10		10		5		1	
1		6		15		20		15		6		1
1	7		21		35		35		21		7	1
1	8	28		56		70		56		28	8	1
1	9	36	84		126		126		84	36	9	1

Now, all we have so far is a pattern that holds for the first eight rows of the table. That may be pretty convincing, but to a mathematician it's only the first step: having observed this pattern in practice, we now want to express it in mathematical terms and see why (and if) it's always true.

Let's start with the mathematical expression. In the examples we cited a moment ago, we said that the sum of the second and third entries on the $n = 4$ row was equal to the third entry of the $n = 5$ row, and that the sum of the third and fourth entries on the $n = 7$ row was equal to the fourth entry of the $n = 8$ row. One way to express this pattern in general would be to say that the kth entry of the nth row— that is, $\binom{n}{k}$—is equal to the sum of the $(k - 1)^{\text{st}}$ and k^{th} entries of the $(n - 1)^{\text{st}}$ row. That is,

$$\binom{n}{k} = \binom{n-1}{k-1} + \binom{n-1}{k}.$$

OK, there's our formula; now: is it true? Actually, it is, and not only that but we have two different ways of seeing it! We can either think of the binomial coefficients as solutions of counting problems, and try to understand in that way why this equation might be true; or we can work with the factorial formula for the binomial coefficients— that is, the formula $\binom{n}{k} = \frac{n!}{k!(n-k)!}$ —and try to manipulate the sum on the left of this equation to see whether it's equal to the quantity on the right.

Let's start with the interpretation of binomial coefficients as solutions of counting problems. We can best describe this approach first by example. Consider for a moment the number of possible bridge hands—that is, collections of 13 cards chosen without repetition from a deck of 52. We know how many such hands there are; it's just the binomial coefficient

$$\binom{52}{13}.$$

But now suppose we divide all bridge hands into two classes: those that include the ace of spades, and those that don't. How many of each kind are there? Well, hands that don't include the ace of spades are just collections of 13 cards chosen without repetition from among the other 51 cards of the deck, so that there are

$$\binom{51}{13}$$

such hands. Similarly, hands that do include ♠A will consist of the ace of spades plus 12 other cards chosen from among the 51 remaining cards of the deck, so that there are

$$\binom{51}{12}$$

of them. Since every hand either does or doesn't contain the ace of spades, the number of all possible bridge hands must equal the number of those that include ♠A plus the number that don't; in other words,

$$\binom{52}{13} = \binom{51}{12} + \binom{51}{13}.$$

The same logic can be applied for any k and n: if we want to count collections of k objects chosen without repetition from a pool of n, we can single out one particular element of our pool of n (it won't matter which one we pick), and break up all collections of k objects from the pool of n into those that do contain the distinguished element, and those that don't. Those that don't, correspond to collections of k objects chosen from among the remaining $n - 1$; those that do, correspond to collections of $k - 1$ objects from among the remaining $n - 1$. Altogether we see that:

> In Pascal's triangle, each binomial coefficient is the sum of the two above it:
> $$\binom{n}{k} = \binom{n-1}{k-1} + \binom{n-1}{k}.$$

There is, as we said, a second way of deriving the formula, by algebra. We know that

$$\binom{n-1}{k-1} = \frac{(n-1)!}{(k-1)!\,(n-k)!} \quad \text{and} \quad \binom{n-1}{k} = \frac{(n-1)!}{k!\,(n-k-1)!}.$$

Now, we're taught in elementary school that if we want to add two fractions the first thing to do is to make their denominators equal. We can do that here: we can multiply the top and bottom of the fraction $\frac{(n-1)!}{(k-1)!(n-k)!}$ by k to arrive at

$$\frac{(n-1)!}{(k-1)!\,(n-k)!} = \frac{k}{k} \cdot \frac{(n-1)!}{(k-1)!\,(n-k)!} = \frac{k \cdot (n-1)!}{k!\,(n-k)!}$$

and likewise we can multiply the numerator and denominator of the fraction $\frac{(n-1)!}{k!(n-k-1)!}$ by $n - k$ to see that

$$\frac{(n-1)!}{k!\,(n-k-1)!} = \frac{n-k}{n-k} \cdot \frac{(n-1)!}{k!\,(n-k-1)!} = \frac{(n-k) \cdot (n-1)!}{k!\,(n-k)!}.$$

Now we can add them: we have

$$\frac{(n-1)!}{(k-1)!\,(n-k)!} + \frac{(n-1)!}{k!\,(n-k-1)!} = \frac{k \cdot (n-1)!}{k!\,(n-k)!} + \frac{(n-k) \cdot (n-1)!}{k!\,(n-k)!}.$$

We can combine the k and the $n - k$ to get

$$= \frac{(k + (n-k)) \cdot (n-1)!}{k!\,(n-k)!}$$

$$= \frac{n \cdot (n-1)!}{k!\,(n-k)!},$$

and merging the n with the $(n - 1)!$ we can rewrite this as

$$= \frac{n!}{k!\,(n-k)!},$$

which we simply recognize as $\binom{n}{k}$. That's the algebraic proof that our formula really does hold!

Exercise 6.1.1. Use our new formula to calculate the 10th and 11th rows of Pascal's triangle from the 9th row.

6.2 PATTERNS IN PASCAL'S TRIANGLE

Here are two more patterns you might observe if you stare at Pascal's triangle long enough.

Suppose first that you decide, on a whim, to add up the binomial coefficients in each row of the triangle. What do you get? The pattern begins to emerge fairly quickly:

$$
\begin{aligned}
1 &= 1, \\
1 + 1 &= 2, \\
1 + 2 + 1 &= 4, \\
1 + 3 + 3 + 1 &= 8, \\
1 + 4 + 6 + 4 + 1 &= 16, \\
1 + 5 + 10 + 10 + 5 + 1 &= 32, \\
1 + 6 + 15 + 20 + 15 + 6 + 1 &= 64.
\end{aligned}
$$

We see, in other words, that the sum of the binomial coefficients in each row is a power of 2; more precisely, the sum of the numbers on the nth row seems to be 2^n.

Why should this be? Well, this is like the last relation, in that once you see it it's not hard to figure out the reason why. There are many ways to think of this, but for fun (and concreteness) let's do the following:

Imagine that we're at a salad bar, which has (say) seven ingredients: lettuce, tomato, onion, cucumber, broccoli, carrots, and the ubiquitous tofu cubes. We ask: *how many different salads can we make?*

Well, this is easy enough to answer using the multiplication principle, if we approach the problem the right way. We can start with a simple choice: should we have lettuce in our salad or not? Next we ask if we want tomatoes or not, and so on; all in all, we have to make seven independent choices, each a yes or no decision. By the multiplication principle, then, we see that there are $2^7 = 128$ possible salads. (Note that we're including the option of just saying no to all the ingredients: the empty salad.)

Suppose on the other hand that we ask: how many different salads can we make that have exactly three ingredients? Again, this is a completely simple application of what we know: there are $\binom{7}{3}$ different ways of choosing three ingredients from among seven. How many salads with two ingredients? $\binom{7}{2}$, of course. With four? $\binom{7}{4}$, and so on.

You can see where this is headed: the total number of salads is the sum of the number of salads with no ingredients, the number of salads with one ingredient, the number of salads with two ingredients, and so on. Since we've already established that the total number is 2^7, we conclude that

$$
\binom{7}{0} + \binom{7}{1} + \binom{7}{2} + \binom{7}{3} + \binom{7}{4} + \binom{7}{5} + \binom{7}{6} + \binom{7}{7} = 2^7.
$$

What's more, you can see that the same idea will work to establish our relation for any n: just imagine a salad bar with n ingredients, and again ask how many salads we can make. On the one hand, the total number is 2^n by the multiplication principle; on the other, it's equal to the sum of the number $\binom{n}{0}$ of salads with no ingredients, the number $\binom{n}{1}$ of salads with 1 ingredient, the number $\binom{n}{2}$ of salads with 2 ingredients, and so on. Thus the sum of all the binomial coefficients with n on top must equal 2^n.

There's an interesting difference, by the way, between this relation and the one we derived in the preceding section. In that case, we had two different ways of verifying the relation: *combinatorially*, that is, via the interpretation of the binomial coefficients as solutions of counting problems; and *algebraically*, that is, by manipulating the factorial formula for binomial coefficients. In the present case, we do have a fairly straightforward combinatorial way of seeing why the formula should be true. But it's not at all clear from the formulas why it holds: if you didn't know that binomial coefficients count collections—if you had only algebra to work with—it would be hard at this point to show that

$$\frac{n!}{0!\,n!} + \frac{n!}{1!\,(n-1)!} + \frac{n!}{2!\,(n-2)!} + \cdots + \frac{n!}{(n-1)!\,1!} + \frac{n!}{n!\,0!} = 2^n.$$

In fact, we will see another way to derive this relation, as well as the following one, from the binomial theorem discussed in the next section.

The next pattern is perhaps more subtle. We're going to look at what's called the *alternating sum* of the binomial coefficients on each row: that is, we're going to look at a particular row and take the first number on that row, minus the second, plus the third, minus the fourth, plus the fifth, and so on to the end. What do we get? Again, the pattern doesn't take long to appear:

$$\begin{aligned}
1 - 1 &= 0, \\
1 - 2 + 1 &= 0, \\
1 - 3 + 3 - 1 &= 0, \\
1 - 4 + 6 - 4 + 1 &= 0, \\
1 - 5 + 10 - 10 + 5 - 1 &= 0, \\
1 - 6 + 15 - 20 + 15 - 6 + 1 &= 0.
\end{aligned}$$

As before, we ask if the alternating sum of the numbers in each row is always going to be 0, and if so why. Note that half the time this is obvious: in the $n = 5$ row, for example, each number appears twice, once with a plus sign and once with a minus, so of course they cancel. The same is true in each row with an odd n. But it's much less clear why this should be true in the even rows as well.

Again, there is a counting-based reason why it should be true. We'll try to phrase it in terms of probability. The key question is: if we flip a coin, say, six times, *what is the probability that we'll get an even number of heads?* Likewise, what is the probability that we'll get an odd number of heads?

Well, first off we can answer these questions in terms of the formulas we've derived in the last few chapters. The probability of getting an even number of heads is simply the sum of the probability of getting no heads, the probability of getting two heads, the probability of getting four heads, and the probability of getting six heads. We've worked all these out in the last chapter; the answer is

$$\frac{\binom{6}{0} + \binom{6}{2} + \binom{6}{4} + \binom{6}{6}}{2^6},$$

and similarly the probability of getting an odd number of heads is

$$\frac{\binom{6}{1} + \binom{6}{3} + \binom{6}{5}}{2^6}.$$

So far, so good. But there's another way to think about the same problem that gets us directly to the answer. It's simply the observation that *it all comes down to the last flip*: whatever the outcome of the first five flips, the question of whether the total number is heads will be odd or even depends on the outcome of the last flip. That is, if the number of heads on the first five flips is even, the total for all six will be even if the last flip is a tail and odd if it's a head; if the number after five flips is odd, the total will be even if the last flip is a head and odd if it's a tail. Either way, the odds are 50-50: the probability of the total number being odd is equal to the probability that the total will be even. But we've already worked out what those odds are, and the conclusion is that the two expressions above must be equal; that is,

$$\binom{6}{0} + \binom{6}{2} + \binom{6}{4} + \binom{6}{6} = \binom{6}{1} + \binom{6}{3} + \binom{6}{5}.$$

But this is exactly to say that the alternating sum

$$\binom{6}{0} - \binom{6}{1} + \binom{6}{2} - \binom{6}{3} + \binom{6}{4} - \binom{6}{5} + \binom{6}{6} = 0,$$

and so we've established our relation in this case. Moreover, it's pretty clear the same logic will work for any n: just imagine we're flipping a coin n times and again ask what the probability is of getting an even number of heads.

We're going to stop here, even though there are many many more patterns to discern in Pascal's triangle. (We'll mention one more in Exercise 6.2.2 below.) It's worth pointing out, though, that these arguments both illustrate a principle that's true in many aspects of life outside mathematics: the key to finding a good answer is to ask the right question.

Here's another way to think about the two relations we've just seen:

Exercise 6.2.1. Use the fact that each number in Pascal's triangle is the sum of the two immediately above it to convince yourself that the alternating sum of the numbers in each row is 0. Can you make a similar argument for the fact that the sum of the numbers in the nth row is 2^n?

Here's one more:

Exercise 6.2.2. Try this: starting with any of the 1s on an edge of Pascal's triangle, and moving along a line parallel to the opposite edge, add up all the numbers until you get to a particular row. For example, if you start with the binomial coefficient $\binom{2}{2} = 1$ and continue to the $n = 6$ row we'd get the sum

$$\binom{2}{2} + \binom{3}{2} + \binom{4}{2} + \binom{5}{2} + \binom{6}{2};$$

or, more graphically, the sum of the boxed numbers in the triangle below

$$
\begin{array}{ccccccccccccccccc}
 & & & & & & & & 1 & & & & & & & & \\
 & & & & & & & 1 & & 1 & & & & & & & \\
 & & & & & & 1 & & 2 & & \boxed{1} & & & & & & \\
 & & & & & 1 & & 3 & & \boxed{3} & & 1 & & & & & \\
 & & & & 1 & & 4 & & \boxed{6} & & 4 & & 1 & & & & \\
 & & & 1 & & 5 & & \boxed{10} & & 10 & & 5 & & 1 & & & \\
 & & 1 & & 6 & & \boxed{15} & & 20 & & 15 & & 6 & & 1 & & \\
 & 1 & & 7 & & 21 & & 35 & & 35 & & 21 & & 7 & & 1 & \\
1 & & 8 & & 28 & & 56 & & 70 & & 56 & & 28 & & 8 & & 1 \\
\end{array}
$$

What do these numbers add up to in this case? In general? Try this a few times and see if you spot the pattern; then see if you can convince yourself that this pattern always holds.

Exercise 6.2.3. Say you flip a fair coin 7 times. What is the probability the number of heads will be even?

6.3   THE BINOMIAL THEOREM

The binomial theorem is concerned with powers of the sum of two numbers. With some trepidation—as you've seen, in this book we try to work with numbers rather than letters when we can—we'll call them x and y. The question is, simply, what do we get when we take the sum $x + y$ and raise it to a power?

Let's start with the first example: let's multiply $x + y$ by itself. (Of course we know you already know how to do this, but bear with us: we want you to think about what you're doing when you do it.) Write it out as:

$$(x + y)(x + y).$$

When we multiply this out, we're going to get four terms: we have x times itself, or x^2; we have y times itself, and we have two cross-terms: an x times a y and a y times a x. Altogether, it adds up to $x^2 + 2xy + y^2$.

Next, consider what happens when we raise $x + y$ to the third power. Again, we'll write it out as a product:

$$(x + y)(x + y)(x + y).$$

Try to anticipate what you're going to get when you multiply this out. First of all, count the number of terms you're going to get: when you expand, you have to pick one of the two terms from each factor $x+y$ and multiply them out; by the multiplication principle there'll be 2^3, or eight, terms. Now, you're going to get one term x^3, the product of the three x's. The next question is: how many terms are you going to get that are a product of two x's and a y? The answer is three: you can choose the x term from any two of the three factors, and the y term from the other. Likewise, you're going to get three terms in the product involving two y's and an x, and one y^3 term; altogether we see that

$$(x + y)^3 = x^3 + 3x^2y + 3xy^2 + y^3.$$

You can probably see where this is headed already, but let's do one more example. Consider the fourth power $(x+y)^4$:

$$(x+y)(x+y)(x+y)(x+y).$$

When we multiply out this product, as before we're going to get one term that is the product of the four x's. Likewise, there are going to be four terms involving three x's and a y: we can pick the x term from three of the factors and the y term from the fourth. And similarly, the number of terms in the expanded product that involve two x's and two y's will be the number of ways of picking two of the four factors in the product: that is, $\binom{4}{2}$, or six. All in all, we see that

$$(x+y)^4 \;=\; x^4 + 4x^3y + 6x^2y^2 + 4xy^3 + y^4.$$

The picture in general is just what you'd expect. When we multiply out the nth power of $x+y$, we get a total of 2^n terms, exactly $\binom{n}{k}$ of which will be the product of k x's and $(n-k)$ y's. What this means is that we have a general formula

$$(x+y)^n \;=\; \binom{n}{0}x^n + \binom{n}{1}x^{n-1}y + \binom{n}{2}x^{n-2}y^2 + \cdots + \binom{n}{n-1}xy^{n-1} + \binom{n}{n}y^n;$$

or, in words:

The coefficient of $x^k y^{n-k}$ in $(x+y)^n$ is $\binom{n}{k}$.

This result is called the binomial theorem, and it's where the binomial coefficients got their name.

One last item: we promised to show you two new ways to see the relations we worked out in the last section. In fact, they're easy to do once we have the binomial theorem.

The point is, the formula for $(x+y)^n$ we just wrote down is an algebraic equation: that is, it's valid if we substitute any two numbers for x and y. So, for example, what happens if we substitute 1 for x and 1 for y? Well, then $x+y=2$, and $x^k y^{n-k} = 1$ no matter what n and k are. So the general formula above now reads

$$2^n \;=\; \binom{n}{0} + \binom{n}{1} + \binom{n}{2} + \cdots + \binom{n}{n-1} + \binom{n}{n},$$

which was our first relation.

Similarly, we could plug in 1 for x and -1 for y. This time the sum $x+y$ is zero, and the products $x^k y^{n-k}$ are alternately 1 and -1: $x^n = 1$, $x^{n-1}y = -1$, $x^{n-2}y^2 = 1$, and so on. Now when we substitute we arrive at

$$0 \;=\; \binom{n}{0} - \binom{n}{1} + \binom{n}{2} - \binom{n}{3} + \cdots.$$

That's our second relation.

Before doing the next two exercises, you might want to review the discussion of multinomial coefficients in Section 4.4.

Exercise 6.3.1. What is the coefficient of $x^3 y^2 z^3$ in $(x + y + z)^8$? (Hint: think about the derivation of the binomial theorem.)

Exercise 6.3.2. What is the sum of all multinomials

$$\binom{7}{a,\ b,\ c}$$

with a 7 on top, and any three numbers adding up to 7 below? (Hint: use your solution to the previous exercise.)

7 Advanced counting

If you've gone through the first five chapters of this book (or even better the first six chapters), you have a pretty good foundation in counting—more than enough, in particular, to see you through the rest of the book. Nonetheless, in the final chapter of this interlude we'd like to take up two additional topics, just to give you an idea of some of what's out there.

Actually, the first of the two topics isn't that much more advanced than what we've already done—it's the formula for the number of collections with repetitions, as discussed in Section 4.5. In fact, the problem is pretty much on a par with the ones we've looked at up to now; it's the derivation, rather than the formula itself, that might be considered less elementary than the content of this book so far.

The second topic—Catalan numbers and some of their applications—is just a really cute exercise in more advanced counting. If you've enjoyed the challenges so far in this part of the book, you may get a kick out of seeing how to tackle a new sort of problem. If not, you could approach it with an anthropologist's point of view: it'll at least give you an idea of some of the more esoteric things mathematicians love to count.

7.1 COLLECTIONS WITH REPETITIONS

Let's start by recalling briefly the discussion in Section 4.5. In the body of this part of the book, we've derived three main formulas.

- We've counted the number of ways of choosing a sequence of k objects (that is, a succession of choices where the order of selection matters) from a common pool of n objects, with repetition allowed: for example, words of a given length, outcomes of k coin flips or dice rolls, and so on.
- We've counted the number of ways of choosing a sequence of k objects from a common pool without repetition: for example, words of a given length with no repeated letters, class officers, etc.
- We've calculated the number of ways of choosing a collection of objects (that is, a succession of choices where the order of selection doesn't matter) from a pool of n objects, without repetition: committees chosen from the pool of students in a class, video rentals, and the like.

If we arrange these in a table, as we did in Section 4.5,

	repetitions allowed	without repetitions
sequences	n^k	$\dfrac{n!}{(n-k)!}$
collections	??	$\dfrac{n!}{k!\,(n-k)!}$

we see that there's an obvious gap: we don't know how to count the number of ways of choosing a collection of objects from a common pool, *with repetition allowed*. It's time to remedy the situation.

Let's start with an example: the fruit bowl introduced in Section 4.5. (We'll change the numbers to make the problem a little more manageable.) The situation is this: you have a fruit bowl, containing (say) eight varieties of fruit—apples, bananas, cantaloupes, durian, elderberries, figs, grapefruit, and honeydews—with unlimited quantities of each. You're assembling a little snack; you figure five servings of fruit would be about right. The question is, how many different snacks can you choose?

Now, if for some reason you decided to rule out taking more than one serving of any one fruit—in other words, if you wanted to consider only snacks consisting of five *different* fruits—this would be a standard problem: we'd be looking at collections of five objects from among eight, without repetition, and the answer would be $\binom{8}{5}$, or 56. But we're not ruling out taking more than one piece of a given fruit; you might decide to go with three apples and two bananas, for example, or this could be a five-banana day. So there should be more possibilities. How many?

Before we go ahead and solve this problem, let's consider some approaches that don't work. For example, we might try the trick we used to count collections without repetition, in Section 4.2: we might first try to count sequences of fruits, rather than collections. (It's a stretch, but imagine we were making a menu for our snack, in which we specified not only the fruits but the order in which they were to be consumed; we could ask how many possible menus there were.) As in Section 4.2, the number of sequences, or menus, is a standard calculation: there are 8^5, or 32,768.

But here the method breaks down. In the case of sequences without repetition, each collection of five objects corresponded to exactly 5!, or 120, different sequences, corresponding to the different orders in which you could eat the five different fruits. But when we allow repetitions, not all collections give rise to 120 different menus: if you went with the 3 apples and 2 bananas, for example, the number of possible menus—the number of ways of ordering the five—would be $\binom{5}{3}$, or 10, corresponding to the number of ways of positioning the apples on your menu. And of course if you decided on the five bananas there would be only one possible menu. So we can't just divide the number of sequences by 120 to arrive at the number of collections, as we did in the no-repetition case.

OK, how about this? We know that there are $\binom{8}{5} = 56$ collections of five fruits without repetition. How many collections are there with two of one fruit and one each of three others? We can do this as well: pick the one fruit we want to double up on—that's eight choices—then choose three fruits without repetition from among the seven fruits remaining—that's $\binom{7}{3} = 35$ choices, for a total of

$$8 \times 35 = 280$$

possibilities. Likewise, we can count the number of collections with three of one fruit and one each of two others (that's $8 \times \binom{7}{2}$, or 168), with two each of two fruits and one of another ($\binom{8}{2} \times 6 = 168$), and so on. We can, in fact, consider in turn all the possible ways of assembling five fruits and add up the number of ways of doing each. But this is a very cumbersome method, and doesn't lead to a simple formula. And, as you might expect, when the numbers get a little larger it becomes completely intractable.

The solution, finally, requires us to approach the problem a little differently: we have to think *graphically* rather than numerically. This is an approach that we haven't seen up to now, but it's something mathematicians do a lot, especially when dealing with counting problems. The general idea is to associate to each of the objects we're trying to count a graphical object—a diagram, a collection of blocks, or whatever—and then count the number of such diagrams that arise. Probably the best way to explain this approach is to illustrate it with examples, and we'll see examples of it both in this section and in the derivation of the formula for the Catalan numbers in Section 7.6.

This sort of approach often requires some ingenuity: it's usually not clear in advance what sort of graphical objects to associate to the things we want to count. In the present circumstance—counting collections of fruits—what we're going to do is to represent each possible choice by a *block diagram*, as we'll describe in a moment. It may not be obvious at first why we're doing this, but (at the risk of mixing food groups) the proof of the pudding is in the eating: at the end, we'll see that we've converted the problem into one we can solve readily.

On to the problem! To begin with, suppose that the fruits are alphabetized: we have apples, bananas, cantaloupes, and so on up to grapefruit and honeydew melons. Now say that we had five white boxes, corresponding to the possible choices of fruits, and seven gray, or "divider," boxes, seven being here one fewer than the number of different fruits we're choosing among. Now suppose we have in mind a particular choice of snack. We can represent that choice by arranging the boxes in a row according to the following rule:

- the number of white boxes to the left of the first divider is the number of apples,
- the number of white boxes between the first and second dividers is the number of bananas,
- the number of white boxes between the second and third dividers is the number of cantaloupes,

and so on, until:

- the number of white boxes between the sixth and seventh dividers is the number of grapefruit, and finally
- the number of white boxes to the right of the last divider is the number of honeydew melons.

Thus, for example, the choice of "an apple, a banana, two figs, and a grapefruit" would correspond to the diagram

while the selection "one cantaloupe, one durian, one fig, and two honeydew melons" would be represented by the diagram

Thus, every selection may be represented by a diagram of 12 boxes, seven of them gray and five white. Conversely, given any such diagram, we can read off from it a choice of fruit: to the diagram

corresponds the choice "one apple, one banana, and three grapefruit."

This construction may have seemed somewhat arbitrary when we first suggested it, but by now the reason should be becoming clearer: what we have is an exact correspondence between ways of choosing a collection of five objects from among eight, with repetition, and box diagrams. The number of collections is thus the same as the number of box diagrams, and this is something we know how to count: it's simply the number of ways of choosing seven boxes from among 12 to color in, and that is

$$\binom{12}{7} = 792.$$

In other words, the number of choices of fruit snacks is the number of possible positions of the seven dividers we need to put in the 12 slots to separate them.

In a similar fashion we can solve the problem of counting collections of k objects from among n in general. We begin by ordering the objects in our pool arbitrarily: we'll call them "object #1," "object #2," and so on up to "object #n." Then, to any such collection, we associate a box diagram, with $n - 1$ gray, or divider boxes, and k white boxes. The rule, as in the fruit example, is simple:

- the number of white boxes to the left of the first divider is the number of times object #1 is chosen,
- the number of white boxes between the first and second dividers is the number of times object #2 is chosen,

and so on until:

- the number of white boxes between the $(n - 2)^{\text{nd}}$ and $(n - 1)^{\text{st}}$ dividers (that is, the last two) is the number of times object #$(n - 1)$ is chosen, and finally
- the number of white boxes to the right of the last, or $(n - 1)^{\text{st}}$, divider is the number of times object #n is chosen.

Just as in our example, we see that in this way we set up a correspondence between collections and box diagrams with k white and $n - 1$ gray boxes—every collection gives rise to a box diagram, and vice versa. And the number of such box diagrams is one we know; it's just the number of ways of saying where the k white boxes should go among the $n + k - 1$ slots in the diagram. The conclusion, in other words, is that:

> The number of ways of choosing a collection of k objects from among n objects, with repetitions allowed, is
>
> $$\binom{n + k - 1}{k} = \frac{(n + k - 1)!}{k!\,(n - 1)!}.$$

We now know how to count both sequences and collections of objects chosen from a common pool, both with and without repetitions. In particular, we can complete the diagram we introduced in Section 4.5:

	repetitions allowed	without repetitions
sequences	n^k	$\dfrac{n!}{(n - k)!}$
collections	$\binom{n+k-1}{k}$	$\binom{n}{k}$

Exercise 7.1.1. Just to do this once: in the fruit example at the beginning of this section, calculate the number of choices consisting of:

1. five different fruits,
2. two of one fruit and one each of three others,
3. two each of two fruits and one of another,
4. three of one fruit and one each of two others,
5. three of one fruit and two of another,
6. four of one fruit and one of another,
7. five of one fruit.

Add these up. Does the result agree with the answer obtained above? (If not, do the problem again.)

Exercise 7.1.2. Suppose again that you're the chief distributor for the Widget Transnational Firm. The Widget Transnational Firm has one central widget-producing plant, and five distribution centers.

Let's say the central plant has just produced 12 crates of widgets, and it's your job to say how many of these 12 each distribution center is to receive. How many ways are there of doing this?

Exercise 7.1.3. By a *rack* in the game of liars Scrabble we'll mean a collection of seven letters; the order doesn't matter, and of course letters can be repeated. How many different racks are possible? (It's not the case in practice, but for the purposes of this calculation, let's say there are at least seven of each letter available.)

Exercise 7.1.4. A company wants to place eight orders for widgets with 13 suppliers.

1. How many ways are there of placing the orders if no supplier is to be given more than one order?

2. How many ways are there of placing the orders if any supplier can be given any number of orders?

3. Of the suppliers, seven are in-state and six are out-of-state. How many ways are there of placing the orders if no supplier is to be given more than one order and at least six orders must be placed in-state?

Exercise 7.1.5. You roll seven identical six-sided dice simultaneously. How many different outcomes are possible? (For instance, one possible outcome is 3 twos, 1 six, 1 five, and 2 ones.)

7.2   CATALAN NUMBERS

Let's start with a strange-sounding question: in how many ways can a collection of parentheses appear in a sentence subject to the restriction you can't close a parenthetical statement that hasn't yet been opened? Say, for example, that a sentence has exactly one pair of parentheses. Ignoring the rest of the text, there's no question: the parentheses must appear in the order "()."

But suppose now that a sentence has two pairs of parentheses. It might have two separate parenthetical statements, so that the parentheses appear like this:

$$()(),$$

or it could have one parenthetical statement included in another, so that if you stripped away all the text the parentheses would be in this order:

$$(()).$$

These are the only possibilities, so we would say there are two ways for two pairs of parentheses to appear in a sentence.

What about three pairs? At this point, you should stop reading and try to work out on your own how many ways three pairs of parentheses might appear in a sentence; we'll tell you the answer on the next page.

The answer is that there are *five* ways they might appear:

$$()()(), \quad ()(()), \quad (())(), \quad (()()), \quad \text{and} \quad ((())).$$

Of course, we're talking mathematics here, not literary style. You won't find too many sentences (in non-technical writing at least) with three pairs of parentheses arranged as in the second of these configurations (and if you did it would (probably) be a fairly convoluted sentence).

For four pairs it turns out there are 14 ways:

$$((())), \quad ((())), \quad (()()), \quad (())(), \quad (()()), \quad (()()), \quad (())(), $$
$$(())(), \quad (())(), \quad (()()), \quad (()()), \quad (()()), \quad (()()()), \quad \text{and} \quad ()()()(.$$

In general, we ask: in how many ways can n pairs of parentheses appear in a sentence? This number is called the nth *Catalan number*, and is denoted c_n. By convention, we say that the 0th Catalan number c_0 is 1. Thus we have seen that

$$c_0 = 1,$$
$$c_1 = 1,$$
$$c_2 = 2,$$
$$c_3 = 5,$$
$$c_4 = 14.$$

The next few Catalan numbers are

$$c_5 = 42,$$
$$c_6 = 132,$$
$$c_7 = 429.$$

The Catalan numbers are a fascinating sequence of numbers that arise in a surprising variety of counting problems. (They were named after Eugène Catalan, a nineteenth-century Belgian mathematician, who encountered them in counting the number of ways you could dissect an n-sided polygon into triangles, a connect we explore in Exercise 7.2.1.) In the remainder of this chapter, we'll mention some of these ways, but the main thing we want to do is to describe a pair of formulas for the Catalan numbers. The first is what we call a *recursive formula*: it expresses each Catalan number in terms of all the previous ones. The second (which is somewhat harder to derive) is by contrast a *closed formula*, which is to say it gives a way of calculating any Catalan number c_n directly; you don't need to know the preceding Catalan numbers to apply it.

Exercise 7.2.1. A *triangulation* of a regular polygon is obtained by drawing chords to connect certain pairs of its vertices in such a way that

- none of the chords cross and
- the resulting interior regions of the polygon are all triangles.

How many ways can you triangulate a square (with vertices labeled)? How many ways can you triangulate a pentagon? How many ways can you triangulate a hexagon? What does this have to do with the Catalan numbers?

7.3   A RECURSION RELATION

The Catalan numbers, as we said, satisfy a *recursion relation*: that is, there is a rule for finding each Catalan number if you know all the preceding ones. This relation is relatively straightforward to see, starting from the description of the Catalan numbers in terms of parentheses.

To derive it, let's go back to the list of the 14 ways that four pairs of parentheses can appear. This time, though, let's try to see if we can list the ways systematically. Of course, any such sequence has to start with a left parenthesis. The key question is: *where in the sequence is its mate?*—that is, the right parenthesis that closes the parenthetical clause it begins. For example, one possibility is that its mate comes right after it: in other words, that the second symbol in the sequence is a right parenthesis, immediately ending the clause begun by the first. What follows then would be simply an arrangement of the remaining three pairs of parentheses, so that the whole sequence would appear as

$$\big(\ \big) \ \{\text{3 pairs}\}.$$

The number of such sequences is just the number of arrangements of three pairs of parentheses, that is, $c_3 = 5$. Explicitly, these are:

$$\big(\big)()(), \quad \big(\big)()(()), \quad \big(\big)(())(), \quad \big(\big)((())), \quad \text{and} \quad \big(\big)((()))\,.$$

Next, it could be that between the initial left parenthesis and its mate there is exactly one pair of parentheses, and after its mate there are two pairs: schematically,

$$\big(\ \{\text{1 pair}\}\ \big) \ \{\text{2 pairs}\}.$$

Now, there is no choice about the arrangement of the single pair between the initial open parenthesis and its mate; but we do have $c_2 = 2$ choices for the last two pairs. There are thus altogether 2 sequences of this type; explicitly,

$$\big(()\big)()() \quad \text{and} \quad \big(()\big)(()).$$

The third possibility is that between the initial left parenthesis and its mate there are exactly two pairs of parentheses, and after its mate there is one pair: schematically,

$$\big(\ \{\text{2 pairs}\}\ \big) \ \{\text{1 pair}\}.$$

As before, there is no choice about the arrangement of the single pair following the initial open parenthesis and its mate; but we do have $c_2 = 2$ choices for the two pairs nested inside them. There are thus altogether 2 sequences of this type; explicitly,

$$\big(()()\big)() \quad \text{and} \quad \big((())\big)().$$

Finally, the last possibility is that the mate of the initial open parentheses is the final right parenthesis: in other words, the entire sequence is part of one parenthetical clause, or schematically,

$$\big(\ \{\text{3 pairs}\}\ \big).$$

There are $c_3 = 5$ such sequences, corresponding to the five possible arrangements of the three pairs of parentheses in the middle; explicitly,

$$\big(()()()\big), \quad \big(()(())\big), \quad \big((())()\big), \quad \big(((()))\big), \quad \text{and} \quad \big(((())))\big).$$

We add up all these possibilities, and we see that $c_4 = 5 + 2 + 2 + 5 = 14$.

We can use the same approach to calculate any Catalan number c_n assuming we know all the ones before it. Basically, we do exactly what we've just done here for c_4: we break up all possible sequences according to where the mate of the initial parenthesis appears—that is, how many pairs appear between them. In other words, every sequence of n pairs of parentheses has schematically the form

$$\Big(\ \{i\ \text{pairs}\}\ \Big)\ \{n - i - 1\ \text{pairs}\}$$

for some i (including $i = 0$ and $i = n - 1$ as possibilities). To specify a sequence of this form we have to choose one of the c_i ways of arranging the i pairs nested inside the initial parenthesis and its mate, and to choose one of the c_{n-i-1} arrangements of the parentheses following. There are thus $c_i \cdot c_{n-i-1}$ sequences of this form, and so we arrive at the recursive formula

---

The nth Catalan number may be computed from the previous Catalan numbers by:

$$c_n = c_0 c_{n-1} + c_1 c_{n-2} + c_2 c_{n-3} + \cdots + c_{n-2} c_1 + c_{n-1} c_0.$$

To express this in words: suppose you write out the Catalan numbers from c_0 to c_{n-1}—for example, if $n = 5$, this would be the sequence

$$1, \quad 1, \quad 2, \quad 5, \quad 14.$$

Now write the same sequence in reverse directly below this one:

$$1, \quad 1, \quad 2, \quad 5, \quad 14,$$
$$14, \quad 5, \quad 2, \quad 1, \quad 1.$$

Now multiply each pair of numbers lying directly over one another:

$$14, \quad 5, \quad 4, \quad 5, \quad 14,$$

and add up these products:

$$14 + 5 + 4 + 5 + 14 = 42.$$

The result will then be the next Catalan number—in this case, $c_5 = 42$. Which you have to admit is a lot simpler than writing out all 42 ways.

Exercise 7.3.1. Check that this rule works for the first four Catalan numbers (starting with c_1), and use it to calculate the next Catalan number c_6.

7.4   ANOTHER INTERPRETATION

Here is another interesting interpretation of the Catalan numbers. Remember when we first introduced binomial coefficients, we described the binomial coefficient $\binom{k+l}{k}$ as the number of possible paths leading from the lower-left to the upper-right corner of a $k \times l$ grid, where at each junction the path went either up (north) or to the right (east).

The reason was, we could describe such a path by a sequence of letters consisting of k Ns and l Es; a sequence such as

E   N   E   E   N   E   N

would correspond to the directions, "go right, then up, then right and right again, then up, then right, and finally up," or in other words to this path:

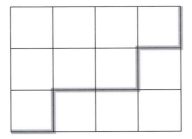

Now imagine that we have 5 pairs of parentheses appearing in a sentence—say

(()()).

Suppose we replace each open parenthesis with an N and each close parenthesis with an E. The resulting sequence

N   N   E   N   E   N   E   E

corresponds to a set of directions taking you from one corner to the opposite corner of a 4×4 grid:

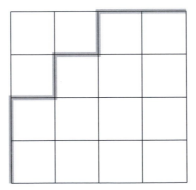

But what does it mean to say that these directions correspond to a "grammatical" sequence of parentheses? If you think about it, it means you can't close a pair of parentheses before you open them—that is, at every point in the sequence you must have had at least as many open (left) parentheses as close (right) parentheses. In terms of the path corresponding to the sequence, this just means that at every stage you must have gone at least as far north as east. In other words, *the path must stay above the diagonal*, though it may touch it. We can thus say that *the nth Catalan number c_n counts the number of paths from one corner of an $n \times n$ triangle to the other*. For example, the number of paths through the grid

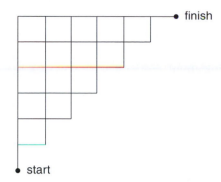

would be the sixth Catalan number c_6.

Exercise 7.4.1. Draw the $c_3 = 5$ possible paths that stay above the diagonal on a 3×3 grid, and the $c_4 = 14$ possible paths that stay above the diagonal on a 4×4 grid.

7.5   THE CLOSED FORMULA

Consider for a moment *all* possible sequences of n left parentheses and n right parentheses, without regard to logic or grammar. We know that the number of such sequences is just the number of ways of placing the n left parentheses among the $2n$ parentheses altogether; that is, it's equal to the binomial coefficient $\binom{2n}{n}$. We might phrase the question: among all ways in which n left parentheses and n right parentheses might appear in a sequence, *what fraction of these ways are grammatically possible?* This suggests an experiment: let's compare the Catalan numbers $1, 1, 2, 5, 14, 42, \ldots$ with the corresponding binomial coefficients, and see if we can see a pattern in their ratios. Here are the results:

n	c_n	$\binom{2n}{n}$	ratio
0	1	1	1:1
1	1	2	1:2
2	2	6	1:3
3	5	20	1:4
4	14	70	1:5
5	42	252	1:6

Well, that's clear enough: based on the evidence so far, it seems natural to think that the nth Catalan number c_n is given by the formula:

The nth Catalan number c_n is given by the formula

$$c_n = \frac{1}{n+1}\binom{2n}{n}.$$

Is it? (Check it for the sixth Catalan number you found in Exercise 7.3.1.) In fact, this formula holds in general, and we'll see why in just a moment. Before we get there, though, you should think about it first on your own: why should this formula hold? It's a fun problem to think about.

One way to express this formula is to say that, of the $\binom{2n}{n}$ paths leading from the lower-left to the upper-right corner of an $n \times n$ grid, the fraction of those that stay above the diagonal is $\frac{1}{n+1}$. If you think in those terms, you'll be able to solve the following problem:

Exercise 7.5.1. Let's say 20 people show up to go see the $5 matinee at the local theater. Suppose that 10 of these people have exact change, but the other 10 have only a $10 bill, and will require change. Unfortunately, on this particular day the cashier has forgotten to stop at the bank and so he has no change to start with. What are the odds that he will be able to sell all 20 people tickets without running out of change?

7.6 THE DERIVATION

We're going to finish this chapter—and this part of the book—by showing you how to count the Catalan numbers from scratch. Like the derivation of the formula for collections with repetitions that we showed you in the first section of this chapter, this will be trickier—less straightforward—than some of the other things we've done so far. Like that derivation, also, it will rely in a fundamental way on a graphical representation of the objects being counted.

Let's start with the interpretation of both Catalan numbers and binomial coefficients as paths through a grid. Suppose, for example, we want to look at paths going five blocks north and five blocks east on a rectangular grid—that is, between the two marked points in the figure below. (We've actually enlarged the displayed grid here, for reasons that will become apparent in moment, but we're still interested just in paths going between the two marked points.)

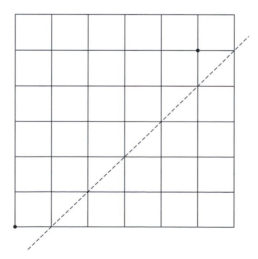

There are $\binom{10}{5} = 252$ such paths altogether; the Catalan number $c_5 = 42$ represents, as we've said, the number of such paths that *do not touch or cross the dotted line*. How do we count such paths? To start with, the first step is to apply the subtraction principle: the number of paths that don't touch the dotted line will be $\binom{10}{5}$ minus the number that do.

Well, that doesn't seem to get us very far: counting the number of paths that do touch the dotted line seems no easier than counting those that don't. Ah, but it is—at least, if we're clever enough. Here's the trick: if we have any path that *does* touch the dotted line, consider the first point at which it touches. We then break up the path into two parts: the first part from the starting point up to that first touch, and the second from that touch to the end, and we *replace the second part of the path with its reflection in the dotted line*. To visualize this, imagine flipping the whole plane over by rotating it 180° around the dotted line, and carrying the second half of the path—that is, the part after the first touch—with it. Thus, for example, if we start with the path

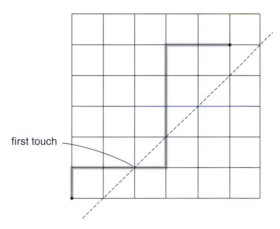

and take the portion of the path after the first touch and reflect in the dotted line, we get

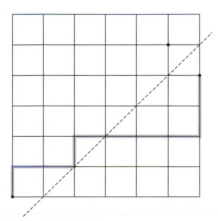

To say it another way: in this example, the original path is represented by the directions

N, E, E, E, N, N, N, N, E, E.

What we are doing is going to the point where the path first touches the dotted line—that would be right after the initial "N, E, E"—and from that point on in the sequence we are switching Ns to Es and Es to Ns to arrive at the sequence

$$N, E, E, N, E, E, E, E, N, N.$$

To put it another way, the path we started with corresponds to the sequence of parentheses

$$( ))) (((( )).$$

What we are doing is going to the point right after the first illegal parenthesis, and from that point on we are reversing each parenthesis; we arrive at

$$( )) ( )))) ((.$$

Here are a few more examples of this process:

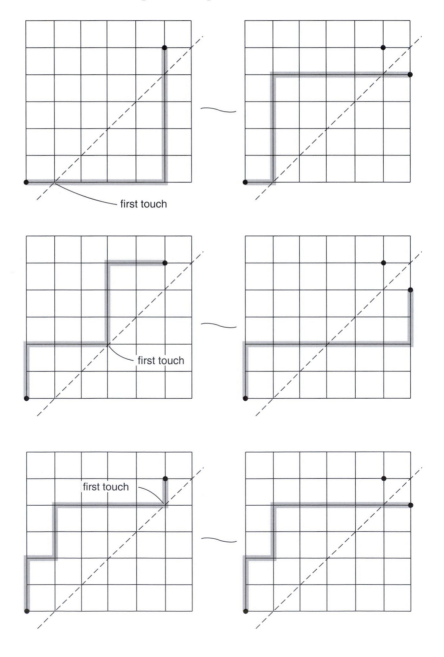

Now, notice one thing about this process. Since the original path goes to the point five blocks north and five blocks east of the starting point, the rejiggered path will go to its mirror image in the dotted line—that is, the point four blocks north and six blocks east of the starting point. Thus, for every path running five blocks north and five blocks east that does touch the dotted line, we arrive at a path going four blocks north and six blocks east.

Moreover, we can always reverse the process. If we have any path going four blocks north and six blocks east, it must cross the dotted line since its destination is on the other side of the dotted line from its origin. We can again look at the point where the path first touches or crosses the dotted line, break the path at that point, and flip the second half of the path around the dotted line. We thus have a correspondence between the two types of paths, and we may conclude that *the number of paths running five blocks north and five blocks east that do touch the dotted line is equal to the number of all paths running four blocks north and six blocks east.*

And that does it: we know the number of all paths running four blocks north and six blocks east is simply the binomial coefficient $\binom{10}{4} = 210$. Thus, of all the $\binom{10}{5} = 252$ paths going five blocks north and five blocks east, 210 do touch the dotted line; it follows that $252 - 210 = 42$ of them don't touch the dotted line, and thus we've calculated the fifth Catalan number $c_5 = 42$.

What's more, exactly the same analysis may be applied to calculate all the Catalan numbers! For any number n, we can draw a similar grid, again with a dotted line one block east of the diagonal. Again, we want to count the paths running n blocks north and n blocks east that stay on or above the diagonal; we do it by counting those that dip below the diagonal (that is, that touch the dotted line) and subtracting this from the total number of paths running n blocks north and n blocks east. And finally, we can count the paths running n blocks north and n blocks east that do touch the dotted line by breaking them at the first touch and reflecting the second part of the path in the dotted line; in exactly the same way we see that *the number of paths running n blocks north and n blocks east that do touch the dotted line is equal to the number of all paths running $n - 1$ blocks north and $n + 1$ blocks east.*

And that's it: the Catalan number c_n—that is, the number of paths running n blocks north and n blocks east that stay on or above the diagonal—is the total number $\binom{2n}{n}$ of paths going n blocks north and n blocks east, minus the number $\binom{2n}{n-1}$ of such paths that touch the dotted line, so that we have a formula

$$c_n = \binom{2n}{n} - \binom{2n}{n-1}.$$

All that remains is to massage this answer a little, to take advantage of common factors in the two terms. Here we go: we've just seen that

$$c_n = \binom{2n}{n} - \binom{2n}{n-1},$$

and expressing these in terms of factorials we have

$$= \frac{(2n)!}{n!\,n!} - \frac{(2n)!}{(n-1)!\,(n+1)!}.$$

Clearing denominators gives us

$$= (n+1) \cdot \frac{(2n)!}{n!\,(n+1)!} - n \cdot \frac{(2n)!}{n!\,(n+1)!}$$

which we can combine to yield

$$= \frac{(2n)!}{n!\,(n+1)!}.$$

Writing the $(n+1)!$ in the denominator as $(n+1) \cdot n!$, we can express this as

$$= \frac{1}{n+1} \cdot \frac{(2n)!}{n!\,n!},$$

or in other words,

$$= \frac{1}{n+1} \binom{2n}{n},$$

and we're done. Our formula is proved, and more to the point the chapter is finished—or almost.

Exercise 7.6.1. Use the closed formula to compute the 8th, 9th, and 10th Catalan numbers.

7.7   WHY DO WE DO THESE THINGS, ANYWAY?

Of what practical value is any of this? This is surely a question that's come up by now, and we (like most writers of math books) are all too aware of it. We're asking you to spend a not-insubstantial amount of time and energy exploring a very esoteric world. What do you stand to gain by it? Why do we do these things, anyway?

There are a number of answers that are traditionally proposed to this sort of question.

Answer #1: To calculate probabilities. In Section 5 we saw how we could use counting skills to estimate probabilities. Surely that's worth something?

Well, yes and no. It's true that counting, of the sort we've introduced here, is the cornerstone of probability theory. But in real life it has pretty limited utility, at least until you get into far more advanced techniques. Let's face it: when you're trying to decide whether or not to call the woman across the table who seems to be saying she has a full house to your flush, it can't hurt to know the odds—but the real question is usually, was that little twitch when she raised inadvertent or deliberate?

Answer #2: To aid in decision-making. Of course knowing how many outfits you can make with four shirts and three pairs of pants doesn't help you actually decide what to wear. But in other situations—for example, distributing the widgets to the warehouses, as in the example at the beginning of this chapter—knowing how many possibilities there are can affect how you go about making the decision among them. It'll give you an idea, for example, of whether it'd be feasible to analyze each possible choice individually, or whether you need some other approach.

This may sound more far-fetched than the first, though it's at least as valid. Again, though, the utility is limited: only in relatively rare circumstances will this sort of calculation be of practical value.

Answer #3: Because mathematics always seems to have applications, even if you don't know what they are when you start out.

The best so far, actually. It really is amazing how even the most abstract mathematics seems to wind up having relevance to the real world far beyond what might have been anticipated. But let's be real: if studying this stuff isn't at least a little bit rewarding to you in itself, the hope of some unspecified application sometime in the future doesn't really cut it as motivation.

But, if we're really being honest, perhaps the real answer is:

Answer #4: It's just plain fascinating.

When you come right down to it, this is not stuff you study in the hopes of a concrete payoff; it's stuff you study just because it's so damn intriguing. Its beauty may not be as immediate as the beauty of music, or art, or literature—you have to dig a little deeper to find it—but it's there. This turns out to be motivation enough for many people— the three of us included—to want to devote our lives to exploring as much of the mathematical universe as we can. And maybe enough motivation for you to continue with us along this journey?

PART II

Probability

8 Expected value

Our discussions of probability thus far have been of a very simple sort. There is an event—a number of coin flips, the roll of a collection of dice, the deal of a hand of cards—with many possible outcomes. We divide the outcomes into two types, which we call "favorable" and "unfavorable." We then ask what is the probability of a favorable outcome; if all outcomes are equally likely—as was the case for the "games of chance" considered in Chapter 5—this is just the number of favorable outcomes divided by the total number of possible outcomes. (The use of words like "favorable" may be misleading: when we ask, for example, for the likelihood that flipping six coins will produce exactly three heads, we're not necessarily rooting for one side or the other; we just want to know the probabilities.)

This model is the heart of most of probability theory. But we need to develop it further, to broaden its range of application. For one thing, in many situations, there are more than two classes of outcomes, each with their own result: think of slot machines or lottery games, which have different payoffs for different types of outcomes. In these situations, we need to be able to assess the total value of a bet—how much we'll win per play, on the average, if we play it repeatedly. This is the notion of *expected value*, which we'll explain in this chapter.

8.1 CHUCK-A-LUCK

As an example of the sort of situation in which the notion of expected value arises, we'll consider some variants of the classic carnival game of Chuck-A-Luck.

The traditional form of Chuck-A-Luck is very simple. To start, you pay a dollar to play. Then you roll three dice; if you roll at least one 6 you win and receive two dollars, so that you win a dollar, net. If not, you get nothing and so lose a dollar net. Presumably, the people who propose the bet are hoping that your intuition runs along the lines of, "well, if there's a one-sixth chance I'll roll a 6 with one die; if there are three dice I should roll at least one 6 half the time, so it's a fair bet."

But this isn't the case. In Section 5.2, we determined the number of possible outcomes when we roll three dice (that is, the number of sequences of three numbers between 1 and 6) is $6^3 = 216$; to calculate the exact probability of rolling at least one 6, we have to count the number of sequences of numbers between 1 and 6 that include at least one 6. We do this by the subtraction principle: the sequences that *don't* include a 6 are simply sequences of three numbers between 1 and 5; there are $5^3 = 125$ of these and so there are $216 - 125 = 91$ sequences with at least one 6.

The probability of winning in Chuck-A-Luck are thus 91/216, or about 42%. That's a pretty hefty profit margin for them, which is why the house can afford to sweeten the deal for you with a new and improved version of Chuck-A-Luck in the hopes of drawing you in. We'll see some ways of doing that, and—more importantly—how you can determine whether or not it's a good bet for you. (Spoiler alert: it's not. No game with the name Chuck-A-Luck is going to be good for you.)

In the new version of Chuck-A-Luck, you still have to pay a dollar to play, and having paid you roll three dice. If none of them is a 6, you receive nothing, and so in effect lose whatever you paid to play. If at least one of them is a 6, you receive two dollars. But if all three come up 6, you receive in addition a grand prize of twenty-five dollars. The question is, what's it worth to play the game? For example, if as before it costs a dollar to play, does that represent a good bet?

To answer this question, it's useful to introduce a new concept, the *expected value* of a bet. To see what this means in this case, suppose we've already paid to play, and we roll the three dice. The point is, there are now three possible classes of outcomes:

- we might roll three 6s, in which case we get $2 + $25 = $27;
- we might roll one or two 6s, in which case we get $2; or
- we might roll no 6s, in which case we get nothing.

How likely are these results? We know how to answer this: for example, as we've seen the first result—three 6s—will occur on average one time per $6^3 = 216$ rolls. Since the payoff for this is $27, your average payoff *on this bet alone* is

$$\frac{1}{216} \times \$27 \sim \$0.125.$$

What about the probability of rolling either one or two 6s? Well, we already saw that there are 91 possible outcomes involving at least one 6; excluding the one outcome consisting of three 6s, we see that there are exactly 90 outcomes involving either one or two 6s. If you play 216 times, in other words, you would expect to receive the $2 prize 90 times; on average, then, the payoff *on this bet alone* is

$$\frac{90}{216} \times \$2 \sim \$0.833.$$

(As always, there is the usual caveat: there is no guarantee that, if you play the game 216 times, you will roll three 6s exactly once and one or two 6s exactly 90 times. What the law of large numbers does say is that, if we continue playing, over time the proportion of the time we roll three 6s will approach 1/216, and the proportion of the time we roll one or two 6s will approach 90/216.)

In any case, adding up the expectations we see that, on average, the payoff per play will be

$$\frac{1}{216} \times \$27 + \frac{90}{216} \times \$2 \sim \$0.958,$$

or about 96 cents. The bottom line, then is that it's worth only about 96 cents to play the game; paying a dollar each time means that in the long run you lose. But, for example, if the carnival offered a special deal of 10 tickets for the price of nine—so that you pay nine dollars to play 10 times, or just 90 cents per play—then the odds would actually favor you.

The average payoff per play, 207/216 dollars, or roughly 96 cents, is called the *expected value* of the play. We will define this concept more formally in Section 8.3, but for now the following slogan is good to keep in mind:

The *expected value* of a game is the average payoff per play.

To get the hang of this, let's try one more variant of this game, Mega-Chuck-A-Luck. In Mega-Chuck-A-Luck you get to roll five dice, with payoffs as follows:

1. if you roll three of a kind—that is, three of the dice come up the same number; it doesn't have to be 6s—you win $3;
2. if you roll four of a kind, you win $10; and
3. if you roll five of a kind, horns are tooted, confetti is thrown and you get a Grand Prize of $100.

The question is, leaving aside the horns and confetti, what's the expected value of the game? In other words, how much, on average, will you win per play, and correspondingly how much is it worth to pay to play? As always, think about this on your own, before we launch into the actual calculation: would this be a good bet if tickets cost three dollars? Two dollars? One dollar?

As before, we have first to calculate the probability of each of the classes of outcomes with payoffs. Start with the easiest: what are the odds of getting five of a kind?

Well, to start, there are $6^5 = 7,776$ possible outcomes, that being the number of sequences of five numbers between 1 and 6. Of these, only six consist of one number repeated five times; thus the odds of getting five of a kind are

$$\frac{6}{7,776} = \frac{1}{1,296} \sim 0.000716.$$

Next, how many of the 7,776 possible outcomes represent exactly four of a kind? To count these, we use the multiplication principle. The number repeated four times can be any number between 1 and 6, for 6 choices; the other number that comes up— remember that we're not including five-of-a-kind outcomes in this count—can be any of the other five numbers between 1 and 6, for 5 choices; and finally this other number can appear in any place in the sequence of five, for 5 choices again. The number of strictly four-of-a-kind outcomes is thus $6 \times 5 \times 5 = 150$, and the probability of this occurring is correspondingly

$$\frac{150}{7,776} \sim 0.01929,$$

or roughly one in 50.

Finally, the odds of three of a kind are calculated similarly: the number repeated three times can be any one of 6 (note that this is unambiguous; you can't have two different numbers each coming up three times in five rolls!); the number of places in which this number can come up in the sequence of 5 is

$$\binom{5}{3} = 10;$$

and the remaining two numbers can be any of the five other numbers between 1 and 6, for $5 \times 5 = 25$ choices. The number of such sequences is thus $6 \times 10 \times 25 = 1{,}500$, and the odds of this occurring are

$$\frac{1{,}500}{7{,}776} \sim 0.1929.$$

Next, we calculate the expected value as follows:

- Six times in 7,776 you'll win the grand prize of \$100; so on average you get

$$\frac{6}{7{,}776} \times 100 = \frac{600}{7{,}776} \sim 0.0716 \text{ dollars}$$

per play. The payoff on this bet alone is thus $\frac{600}{7{,}776}$ dollars, or roughly 7 cents per play.
- 150 times in 7,776 you'll win \$10; so on average the payoff on this bet alone is

$$\frac{150}{7{,}776} \times 10 = \frac{1{,}500}{7{,}776} \sim 0.1929 \text{ dollars},$$

or roughly 19 cents per play.
- Finally, 1,500 times in 7,776 you'll receive three dollars; so on average you get

$$\frac{1{,}500}{7{,}776} \times 3 = \frac{4{,}500}{7{,}776} \sim 0.5787 \text{ dollars},$$

or roughly 58 cents per play.

The expected value of a play of Mega-Chuck-A-Luck is thus

$$\frac{6}{7{,}776} \times 100 \; + \; \frac{150}{7{,}776} \times 10 \; + \; \frac{1{,}500}{7{,}776} \times 3$$
$$= \frac{6{,}600}{7{,}776}$$
$$\sim 0.8487 \text{ dollars},$$

or approximately 85 cents per play. In other words, it's a *terrible* bet if tickets cost a dollar.

If we could detour for a moment from probability to psychology, we should point out that this sort of payoff scheme is typical of gambling games: the relatively high-payoff, low-probability outcomes get the attention (you can bet the signs outside the booth are screaming, "Play Mega-Chuck-A-Luck! Win a Hundred Dollars!," not, "Play Mega-Chuck-A-Luck! Win Three Dollars with Significant Likelihood!"). Many bettors tend to focus on them; but it's the payoffs at the other end of the spectrum that typically matter more. To illustrate this, consider the following example:

Problem 8.1.1. Let's consider two ways in which we might bump up the payoffs in Mega-Chuck-A-Luck.

- Suppose we sweeten the Grand Prize (for five of a kind), making it 250 dollars rather than 100. What's the expected value now? Is it worth a buck to play?
- On the other hand, suppose we make the prize for three of a kind 4 dollars rather than 3, and leave the other payouts the same at 10 and 100 dollars. Again, what's the expected value?

Solution. We already know the probabilities of the relevant outcomes (three, four, and five of a kind), so this is pretty simple. For the first part, we have to replace the 100 by 250 in the calculation we just did; the expected value of this version of Mega-Chuck-A-Luck is

$$\frac{6}{7,776} \times 250 + \frac{150}{7,776} \times 10 + \frac{1,500}{7,776} \times 3$$
$$= \frac{6,600}{7,776}$$
$$= .179 + .1929 + .5787$$
$$\sim 0.9506.$$

Similarly, for the second part we have to replace the payoff for three of a kind by 4; the expected value of this version of Mega-Chuck-A-Luck is

$$\frac{6}{7,776} \times 100 + \frac{150}{7,776} \times 10 + \frac{1,500}{7,776} \times 4$$
$$= \frac{6,600}{7,776}$$
$$= .0716 + .1929 + .7716$$
$$\sim 1.036.$$

The second variation, while less flashy, is much better for us. □

So what have we learned? Rephrasing our analysis in terms of expected value (instead of just the probability of winning versus the probability of losing, as we did in the analysis of the classic version of Chuck-A-Luck) does two things for us:

- it allows us to analyze a much broader range of games than before: we can now analyze games with multiple classes of outcomes and correspondingly many possible results; and
- by determining quantitatively the value of a play, we can adjust our decision if the price of a play changes, as in the example of the carnival offering a 10-for-9 deal.

We'll see below many other examples of games to which we can apply this analysis. As usual, we're discussing the concept of expected value primarily in the context of gambling games because this sort of artificial environment allows us to assign exact probabilities to various events and to calculate them; but it's an idea with a very broad range of applicability, as we'll also try to illustrate.

Exercise 8.1.2. Suppose that a carnival is selling tickets, each of which allows you to play Mega-Chuck-A-Luck (as described above) once. As we saw, if the price per ticket is one dollar, it's not a good bet, while if they're offering 10 tickets for nine dollars the odds are in your favor.

1. Suppose the carnival is offering 25 tickets for 24 dollars. Do you go for it?
2. How about if they offer 20 tickets for 19 dollars?

Exercise 8.1.3. In their never-ending quest to siphon the money from your wallet, the carnival operators have come up with a new game: Super-Mega-Chuck-A-Luck! Here you roll seven dice, with payoffs as follows:

- if you roll five of a kind, you receive 50 dollars;
- if you roll six of a kind, you receive 500 dollars; and
- if you roll seven of a kind, you get a Grand Prize of 5,000 dollars.

What's the expected value of Super-Mega-Chuck-A-Luck?

We promise, this is the last time we'll mention Chuck-A-Luck in any form. You may, however, notice a distinct similarity between it and the lottery games described below. This is distressing, at least if you'd like to think of your state government as something other than a bunch of carnies.

8.2   WHY WE'RE SPENDING SO MUCH TIME AT THE CASINO

We said that you can apply the notion of expected value in situations other than gambling. Let's consider an example of this:

Problem 8.2.1. Ira and his friends are driving to a movie. When they're a mile away from the theater, they see a parking space; if they park there it'll take them 20 minutes to walk to the movie. If they do look for a closer space, there's:

- a 40 percent chance they'll find a space a half-mile from the theater, a 10-minute walk;
- a 40 percent chance they'll have to come back and take the original space; and
- a 20 percent chance that when they do come back the space will be gone and they'll have to park a mile-and-a-half away, a 30-minute walk.

Ira calculates the average amount of time it'll take them to get to the theater if they do look for a better spot, and proposes that they do this, as on average it'll take them less than 20 minutes. His friends disagree. Who's right?

Solution. The problem asks us to find the expected value of the time it'll take Ira and his friends to get to the theater, if they adopt Ira's suggestion of looking for a better spot, and to compare it to the 20 minutes it'll take them if they just park where they are. To do this, suppose Ira and his friends find themselves in this situation five times, and each time opt to look for a closer space. On average:

- twice they'll find the better space, and take 10 minutes to get to the theater;
- twice they'll wind up in the original spot, and take 20 minutes; and
- once they'll have to park a mile-and-a-half away, and take 30 minutes.

The total amount of time they'll take, then is

$$2 \times 10 + 2 \times 20 + 1 \times 30 = 90 \text{ minutes,}$$

for an average of $90/5 = 18$ minutes. Ira was right! ... Or was he?

Now might be a good time to pause and ask ourselves what's wrong with this picture? The parking problem illustrates how the notion of expected value can be applied in a variety of situations. It also illustrates many of the things that are wrong with word problems purporting to do this. To wit:

- The basic premise—that there are just three possible outcomes of hunting for a parking spot—is clearly bogus.
- The idea that we can assign meaningful probabilities to these outcomes is also unrealistic.
- Real life is messier than that: for example, we haven't taken into account the time it takes to drive to the better parking space, and (possibly) back.
- The idea that you can always quantify outcomes is suspect. In this case, if the movie starts in 22 minutes, taking the parking space a mile away assures you of getting there on time; finding a better spot would only ensure that you had to sit through a bunch of advertisements. If the movie started in 12 minutes, though, it might make more sense to gamble.

Frankly, many of the word problems purporting to apply concepts like expected value to the real world suffer from one or more of these defects. Which is why we're spending so much time at the casino: at least there the outcomes are finite, the probabilities precise, and the result—money won or lost—appropriately quantifiable.

Exercise 8.2.2. Devise your own "real-world" expected value word problem and then enlist a friend to critique the applicability of your expected value calculation. What are the dangers of computing the expected amount of time it will take you to get to the gate for your next flight? Would it change your willingness to play a casino game with a positive expected value if you need some of the money you might use to play to pay for dinner?

8.3   EXPECTED VALUE

It's time to formalize the notion of expected value; then we'll work out some more examples.

To start, we suppose that we perform an experiment, which will have one of two possible results; we'll call these Result P and Result Q. Suppose moreover that we can assess the probability of Result P occurring, which we'll call p, and the likelihood of Result Q occurring, which we'll call q. (The numbers p and q represent the fraction of the time Results P and Q will occur, if the experiment is repeated over and over. In particular, since we assumed that one or the other must occur, we have $p + q = 1$.)

Suppose finally that we have associated to each possible result a "payoff;" we'll say the payoffs associated to Results P and Q are a and b respectively. (Again, in gambling situations, this is the amount of money won if the relevant event occurs, but in general it doesn't have to be money, and it doesn't—despite the name "payoff"—have to be something desirable. In the parking problem, for example, it's the time it takes to get to the theater.) In this case, we define the *expected value* of the experiment to be the sum

$$\mathrm{ev} = p \cdot a + q \cdot b.$$

The significance of this expression should be clear from the examples we've worked out already. If we were to carry out the experiment a large number N of times, we would expect Result P to occur pN times, resulting in a total payoff of pNa on those occasions. Likewise, Result Q will occur qN times, resulting in a total payoff of qNb; the total payoff for the N plays is thus $pNa + qNb$, and the *average* payoff per iteration of the experiment is

$$\text{ev} = \frac{pNa + qNb}{N} = pa + qb,$$

and this is what we call the expected value.

In these terms, it's straightforward to extend this to the case where there are more than two possible results. Suppose in fact that there are k possible results, which we'll call $P_1, \ldots, P_k$; suppose that the probability of Result P_i occurring is p_i, and that the payoff in that case is a_i. In that case, we define the expected value of the experiment to be the sum

$$\text{ev} = p_1 a_1 + p_2 a_2 + \cdots + p_k a_k.$$

This gives a general formula for expected value:

If an experiment has k possible results, occurring with probabilities $p_1, \ldots, p_k$ and payoffs $a_1, \ldots, a_k$, the *expected value* of the experiment is

$$\text{ev} = p_1 a_1 + p_2 a_2 + \cdots + p_k a_k.$$

Problem 8.3.1. Suppose you roll one die, and are given as many dollars as the number showing—that is, if you roll a 1, you get one dollar, and so on. What's the expected value?

Solution. Here there are six possible results, each of which will occur with probability $1/6$. The average payoff—the expected value, in other words—will be

$$\begin{aligned}
\text{ev} &= \frac{1}{6} \cdot 1 + \frac{1}{6} \cdot 2 + \frac{1}{6} \cdot 3 + \frac{1}{6} \cdot 4 + \frac{1}{6} \cdot 5 + \frac{1}{6} \cdot 6 \\
&= \frac{21}{6} \\
&= 3.5.
\end{aligned}$$

Thus, it's worth three and a half dollars to play.

Exercise 8.3.2. Suppose you roll two dice and the payoff for the roll is the sum of the numbers appearing on the dice. What is the expected value of the game?

Exercise 8.3.3. In a drastically simplified poker game, you're dealt just two cards. If you get a pair—that is, if the two cards are of the same denomination—you receive $50. If you get a "flush"—two cards of the same suit—the payoff is $10. What's the expected value of the game?

8.4   STRATEGIZING

We mentioned above that one of the tricks the gambling industry[1] uses to mislead you about the actual expected value of a bet is to focus on the high-payoff, low-probability end of the spectrum of possible events. There's another standard technique: to have so many different scenarios, each with its own payoff and probability, that it becomes too daunting to work out the expected value. This is the case especially with slot machines, which started out as relatively simple mechanical devices but evolved into enormously complicated machines driven by electronics. Here, we're aiming just to illustrate the notion of expected value, and not to give you advice on how to play actual slots. (That advice, if you're interested, is simple: don't.) So we'll work with a vastly simplified version of slot machines, akin to the earlier models.

Here's the setup. The slot machine in question has three reels, each of which has five pictures on it: say, apples, cherries, lemons, grapes, and a bell. When you pull the lever, the reels rotate, each one coming to a stop on one of the five pictures; we'll assume (though we notice that most slot machines don't actually say this) that which picture comes up on each reel is random, that all pictures are equally likely, and that they're independent—which picture came up on the first reel doesn't affect which picture will come up on the second. Here are the payoffs:

- three bells pays $50;
- three of any other symbol pays $10;
- two bells, plus any other picture, pays $2.

What's the expected value? If it costs a dollar to play, is it worth it?

Again, the first step is to calculate the probability associated to each result. To begin with, there are $5^3 = 125$ possible outcomes, all (purportedly) equally likely. There's only one way of getting three bells, so the odds of that happening are just 1 in 125. Similarly, there's just one way of getting three of any picture, so the probability of getting three of any of the other four pictures is 4 in 125.

As for the likelihood of getting two bells and one of another picture, we can count the number of the 125 outcomes of this type: first we have to say which two of the three rolls stop on the bell, which is $\binom{3}{2} = 3$ choices; then we have to say what the other symbol is, which is 4 choices. Thus, 12 of the 125 outcomes are of this type, and the odds of this happening are 12 in 125.

Next, to calculate the expected value we just multiply the likelihood of each event by the corresponding payoff, and add up: we have

$$
\begin{aligned}
\text{ev} &= \frac{1}{125} \times 50 \; + \; \frac{4}{125} \times 10 \; + \; \frac{12}{125} \times 2 \\
&= \frac{50 + 40 + 24}{125} \\
&= \frac{114}{125}.
\end{aligned}
$$

---

[1] Excuse us; that should be the *gaming entertainment industry*, which, we are assured by the website of the American Gaming Association, offers some of the most dynamic and rewarding professional opportunities available today.

So the answer to the question, "If it costs a dollar to play, is it a good bet?" is no. (We may be simplifying, but we do try to be at least a little realistic.)

Being able to calculate expected values gives us a strategic tool: confronted with a choice of several possible bets, we can calculate the expected value of each to determine our optimal strategy. We'll illustrate this with a couple easy examples, and then go on to consider a less cut-and-dried situation.

Say you have a choice of three games. In each, you are simply dealt one card at random from a standard deck. The payoffs, though, are different:

1. Game A is the simplest: if your card is an ace, you win $100; otherwise, you get nothing.
2. In Game B, you win $50 if you get an ace, and $25 if you get a face card (that is, a jack, queen, or king); otherwise, you get nothing.
3. In Game C, you win $25 if you get a face card, and otherwise you receive the number of dollars showing on your card: $1 for an ace, $2 for a 2, and so on up to $10 for a 10.

Which game is the most favorable to you, and how much is it worth to play?

Answering this—that is, calculating the expected value of each game—is straightforward. Game A in particular is easy: you win $100 one-thirteenth of the time, so the expected value in dollars is

$$\frac{1}{13} \cdot 100 = \frac{100}{13} \sim \$7.69.$$

As for Game B, you win $50 one-thirteenth of the time, and $25 another three-thirteenths of the time; the expected value is thus

$$\frac{1}{13} \cdot 50 + \frac{3}{13} \cdot 25 = \frac{50 + 3 \cdot 25}{13} = \frac{125}{13} \sim \$9.62.$$

Finally, for Game C the expected value is

$$\frac{1}{13} \cdot 1 + \frac{1}{13} \cdot 2 + \cdots + \frac{1}{13} \cdot 9 + \frac{1}{13} \cdot 10 + \frac{3}{13} \cdot 25 = \frac{1 + 2 + \cdots + 9 + 10 + 75}{13}$$
$$= \frac{130}{13}$$
$$= \$10.$$

So Game C is the most favorable, and worth $10 to play.

Let's work through a less cut-and-dried example. Here's the deal: to start, you pay $3,000. Then you get to roll a bunch of (standard, six-sided) dice. The results are as follows:

- If you roll at least one 6, but no 1s, you get a payoff of $10,000.
- Otherwise—if you either roll a 1, or fail to roll a 6—you get nothing.

Here's the wrinkle: *you get to decide how many dice to roll.* So the question is, what number of dice gives you the best expected value?

The first thing to see when you start thinking about this is that, for any given number of dice, we can calculate the expected value. We'll do this in the cases of one, two and three dice, and then, when we've seen the pattern, work out the expected value for any number.

First, for one die, it's very simple: there's a 1 in 6 chance you'll roll a 6, and get the prize; otherwise, you get nothing. So the expected value is

$$\frac{1}{6} \times 10{,}000 = \frac{10{,}000}{6} \sim 1{,}667 \text{ dollars.}$$

What about two dice? Again, we have to determine the number of outcomes (sequences of two numbers between 1 and 6) that involve at least one 6, but no 1s. There are basically two kinds of such outcomes: we could roll two 6s, or one 6 and one of another number between 2 and 5. There's just one way to do the former, and $\binom{2}{1} \times 4 = 8$ ways to do the latter, for a total of 9. Since there are 36 equally likely possible outcomes, the expected value is then

$$\frac{9}{36} \times 10{,}000 = \frac{10{,}000}{4} = 2{,}500 \text{ dollars.}$$

Better! Now, how about three dice? Here there are three ways of rolling at least one 6 but no 1s: all three 6s, which is one possible outcome; two 6s and one other number between 2 and 5, which represents $\binom{3}{2} \times 4 = 12$ possible outcomes; and one 6 with two others between 2 and 5, which can occur in

$$\binom{3}{1} \times 4^2 = 48$$

ways. Altogether, there are $1 + 12 + 48 = 61$ such rolls, and the expected value is

$$\frac{61}{216} \times 10{,}000 = \frac{610{,}000}{216} \sim 2{,}824 \text{ dollars.}$$

Better still! At this point, though, the croupier's getting subtly impatient, the people behind us are clearing their throats, and we'd better decide soon; so let's cut to the chase and calculate the expected value ev_n if we roll any number n of dice.

What we need is a better way to calculate these numbers in general. Happily, we have one: the subtraction principle. Suppose we roll n dice. There are 6^n possible outcomes. Of these, we know how many have no 1s: that would just be the number of sequences of n numbers between 2 and 6, or 5^n. Now we ask: among those 5^n outcomes that have no 1s, how many have at least one 6? The subtraction principle gives us the answer: the number of those have *no* 6s is just the number of sequences of n numbers between 2 and 5, or 4^n, so the number that have no 1s and at least one 6 is the difference, $5^n - 4^n$. We can thus write the expected value ev_n in general as

$$\text{ev}_n = \frac{5^n - 4^n}{6^n} \times 10{,}000.$$

OK! We whip out our calculator and make a table of what ev_n is for different choices of n:

n	ev_n	whole-number approximation
1	$10{,}000/6$	1,667
2	$90{,}000/36$	2,500
3	$610{,}000/216$	2,824
4	$3{,}690{,}000/1{,}296$	2,847
5	$21{,}010{,}000/7{,}776$	2,702
6	$115{,}290{,}000/46{,}656$	2,471
7	$617{,}410{,}000/279{,}936$	2,206

and ev_n just gets smaller and smaller thereafter. The answer is clear: ignoring the by-now openly hostile croupier and the approaching floor manager, you say with firm decision, "I'll roll four dice!".

And prepare to lose your shirt. We said that rolling four dice gave you the best expected value, but we didn't say it was actually a good bet: you're paying \$3,000 to play a game worth \$2,847.

Problem 8.4.1. Suppose we modify the game in your favor: now, you win if you get *either* a 5 or a 6, but no 1s; otherwise, you lose. What's the optimal number of dice to roll in this situation, and what expected value do you have now?

Solution. The first step has to be to calculate the probability of winning this version of the game if you roll n dice. We can do this pretty much the same way as above. Again, of the 6^n possible outcomes, there are 5^n that involve no 1s. Of these, the number that have no 5s or 6s is 3^n, so the number of winning rolls is now $5^n - 3^n$, and the expected value is

$$\mathrm{ev}_n = \frac{5^n - 3^n}{6^n} \times 10{,}000.$$

Once more, we make a table of these values for $n = 1, 2, 3, \ldots$:

n	ev_n	whole-number approximation
1	$10{,}000 \cdot 2/6$	3,333
2	$10{,}000 \cdot 16/36$	4,444
3	$10{,}000 \cdot 98/216$	4,537
4	$10{,}000 \cdot 544/1296$	4,197
5	$10{,}000 \cdot 2882/7776$	3,706

and ev_n just gets smaller and smaller thereafter. So the optimal strategy here is to roll 3 dice.   □

We'll let you work through one more variant:

Exercise 8.4.2. Suppose we modify the game in the house's favor, so that you win if you get a 6, but no 1s *or* 2s; otherwise, you lose. Again, what's the optimal number of dice to roll, and what's your expected value?

Exercise 8.4.3. Here are two more games, played along the same lines as Games *A*, *B*, and *C* above. In each case, you get one card at random, with payoffs as follows:

d. In Game D, you win $10 if you get a face card, and otherwise the payoff is twice the number showing on your card: $2 for an ace, $4 for a 2, and so on up to $20 for a 10.
e. In Game E, you get $20 for a face card; if your card is a number card (we count an ace as a 1) you get $10 if the number is even and $5 if it's odd.

Compare these games to Games A, B, and C above.

Exercise 8.4.4. Consider a slot machine like the one described above, with three reels, each of which has five pictures: apples, cherries, lemons, grapes, and a bell.

1. Suppose, however, that the payoffs are different: three bells pays $25, three of any fruit pays $10, and two bells plus a fruit pays $3. What's the expected value of this game?
2. Suppose the payoffs have changed to $50 for three bells and $20 for any three fruits (that is, no bells). Now what's the expected value?
3. After playing slots for several hours, you are shocked to realize that the machine actually has six pictures on each reel: one bell and five different fruits. If the payoffs are $50 for three bells, $10 for three of any fruit, and $2 for two bells and a fruit, what is the expected value now?

8.5   MEDICAL DECISION-MAKING

There is one more very important example of strategizing that is worth discussing explicitly before we move on, this time in a medical setting. Let's say at the outset that the situation and numbers here are grossly oversimplified and unrealistic; we're just trying to illustrate the concept.

Suppose Allie, who is 60, has a life-threatening illness, and a choice of two surgical procedures to address the problem. Procedure A has the highest chance of an outright cure, but is also the riskiest. In 40% of cases similar to hers the procedure fixed the problem outright, so that the patient lived out, on average, their life expectancy, which in Allie's case is 20 years. In 20% of similar cases, however, the patient either died in surgery or from complications; the remaining 40% died an average of five years later from a recurrence of the problem. Procedure B offers a smaller chance of outright cure, but is also safer: only 20% are cured outright, but no one dies; the remaining 80% live an average of 5 years. You can guess the question: which procedure offers the greater expected value for number of years?

It's a simple application of our formula: Procedure A has an expected value of

$$\frac{40}{100} \times 20 + \frac{40}{100} \times 5 = 10 \text{ years,}$$

while Procedure B has expected value

$$\frac{20}{100} \times 20 + \frac{80}{100} \times 5 = 8 \text{ years.}$$

So Procedure A will, on average, give Allie more years to live.

Now, in Section 8.2, we gave a similarly made-up problem involving parking. We then proceeded, in Section 8.2, to point out how bogus the problem was (somewhat disingenuously, we admit, since we made up the problem in the first place). In fact, similar objections can be made here, namely:

1. The body's reaction to surgery varies in a spectrum; it can't be categorized simply as a success or a failure.

2. Assigning probabilities to the various results is questionable. These are usually based on prior experiences of people similar to you. But similar in what ways? We don't necessarily know what external or internal factors affect the success of the operation.

3. Quantifying the outcome is bogus as well. Is 20 years' life expectancy four times better than five? For that matter, is (say) a one-in-four chance of living 20 years the equivalent of a certainty of living five?

Again, all these objections are legitimate. But there's a difference: in the parking problem the decision is ultimately unnecessary; Ira's friends can (and probably will) mock him mercilessly until he gives up and just parks. But medical decisions really do have to be made one way or another, and it's important to understand the basis on which they're made, flawed as it may be.

This, ultimately, is a large part of what we hope you'll take away from the course. Of course we want to you to be able to determine probabilities in those circumstances where they're both meaningful and calculable; that has to be the foundation of everything that follows. But there are also many situations out there where the application of this sort of probabilistic reasoning is clearly flawed—and has to be done anyway. In those cases a clear sense of the extent to which probability is and isn't applicable is vital. We'll talk a bit more about some of these issues when we discuss large-scale applications of probability theory and in our concluding remarks in Chapter 14.

Exercise 8.5.1. Suppose now that Allie is actually 70 years old, and that her life expectancy (if cured) is 12 years rather than 20. Which procedure now has the greater expected value? What if her life expectancy is 8 years?

9 Conditional probability

In Chapters 5 and 8 we computed probabilities associated to *independent* events, like coin flips or dice rolls, in which the results of each trial had no effect on the results of subsequent trials. But life's not always like that. In this chapter we'll introduce *conditional probability*, where the likelihood of a given outcome can be affected by previous events. We'll consider cases where two outcomes are *positively correlated*, meaning that one is more likely to occur when the other is known to have occurred, and also cases where two outcomes are *negatively correlated*. A highlight will be *Bayes' theorem*, a key tool in computing the conditional probability of two correlated events.

9.1 THE MONTY HALL PROBLEM

The Monty Hall problem refers to a scenario that arose in every iteration of an old game show called "Let's Make a Deal." The setup varied in its particulars, but never in its general outline, which went like this: first, the host, Monty Hall, would select an audience member to play the game. The player would be shown three curtains, behind one of which was a desirable prize—a naugahyde living room set, for example, or a car—and behind the other two there were joke prizes, clearly viewed as worthless, like a goat.[1] The player would get to choose one of the three curtains, and could receive the prize behind that one.

At this point, however, before revealing the prize behind the curtain selected by the player, Monty would open one of the curtains not chosen, and reveal that behind it was a goat. The player would then have the option of sticking with their original choice, or switching. The questions are: should the player switch, and what are the probabilities of success in each case?

(Note that Monty does not choose randomly between the two doors that the player didn't select and show what's behind that one. He always picks a door with a goat. So, if the player's initial guess was a door with a goat, Monty has no choice at all of which door to open; he has to pick the other goat door. But if the player's original door was the one with the car, Monty can pick either one of the two remaining doors to open; he does so randomly.)

The first thing to point out is that if you choose to stick with the original choice, your likelihood of winning are the same as they were before Monty got involved: one

---

[1] We like goats. If you come right down to it, we like them more than naugahyde living room sets. But for the purpose of this problem, we'll go along with Monty and deem the latter desirable, and the former not.

in three. On the other hand, to figure the probability if you choose to switch, we can reason that if you switch:

- One time in three your original guess will have been the right one; in those cases you now lose. But:
- Two times in three your original guess will have been wrong; in these cases, you win.

Thus your probability of winning if you switch is 2 out of 3!

Let's try a few variations on this game, and see where it leads us. For example, what if there are four doors, with one car and three goats behind them. We play the same game—we pick a door, Monty shows us a goat, we choose to stick or switch—but suppose now there are four doors. What are the probabilities either way in this case?

As before, if we stick we'll win one in four times. What if instead we decide we'll switch to one of the two remaining doors? In that case:

- One time in four your original guess will be the right one. In those cases you now lose. But:
- Three times in four, your original guess will have been wrong. In these cases, the winning door is one of the remaining two and you'll guess right half the time.

Put it this way: you'll be right one-half of three-quarters of the time—in other words, three times in eight. Again, better than sticking.

At this point, we might as well go all the way and do the case of n doors, again with one car and $n-1$ goats. If we stick, the probability we win is $1/n$, as before. If we switch, the logic goes:

- One time in n your original guess will be the right one. In those cases you now lose. But:
- $n-1$ times in n, your original guess will have been wrong. In these cases, the winning door is one of the remaining $n-2$ and you'll guess right $1/(n-2)$ of the time.

The probability that you win is thus

$$\frac{n-1}{n} \cdot \frac{1}{n-2} = \frac{n-1}{n(n-2)},$$

which we can write as

$$\frac{n-1}{n(n-2)} = \frac{1}{n} \cdot \frac{n-1}{n-2}.$$

Since $n-1/n-2 > 1$, we see that this is always better than the probability $1/n$ of winning we get if we stick with our original choice.

We have to ask:[2] what if there are multiple doors, and multiple cars? For example, suppose there are five doors, with two cars and three goats behind them—stick or switch? The same logic applies: if you stick, the probability you win is simply two in five. On the other hand, if you switch:

- Two times in five, your original guess will have been one of the right ones. In this case, after Monty shows you a goat, the remaining three doors will have behind them one car and two goats; in this case, you win one-third of the time.

---

[2]   Really, we do. We're like that.

- Three times in five, your original guess will have been one of the wrong ones. In these cases, the remaining three doors will have two cars and one goat behind them; in this case, you win two-thirds of the time.

In other words, 2/5 of the time you'll have a one-in-three chance of winning; 3/5 of the time you'll have a two-in-three chance of winning. The probability of your winning is thus

$$\frac{2}{5} \cdot \frac{1}{3} + \frac{3}{5} \cdot \frac{2}{3} = \frac{8}{15}.$$

At this point, we might as well go all the way and do the case of an arbitrary number n of doors, and an arbitrary number k of cars. (Well, not completely arbitrary: there have to be at least three doors, or we can't play the game, and likewise there have to be at least two goats, so that no matter which door you pick Monty can show you a goat; in other words, $n \geq 3$ and $k \leq n - 2$.) In this case, the probability that we win if we choose to stick is simply k/n. On the other hand, if we choose to switch, we have the by-now familiar logic:

- k/n of the time, your original guess will have been one of the right ones. In this case, after Monty shows you a goat, the remaining $n - 2$ doors will have behind them $k - 1$ cars and $n - k - 1$ goats; your chances of guessing right are correspondingly $\frac{k-1}{n-2}$. On the other hand:
- $(n-k)/n$ of the time, your original guess will have been one of the wrong ones. In these cases, the remaining $n - 2$ doors will have behind them k cars and $n - k - 2$ goats; your chances of guessing right are now $\frac{k}{n-2}$.

Adding up, the probability that you win is

$$\frac{k}{n} \cdot \frac{k-1}{n-2} + \frac{n-k}{n} \cdot \frac{k}{n-2} = \frac{k(k-1) + (n-k)k}{n(n-2)}$$
$$= \frac{k(n-1)}{n(n-2)}.$$

Note that this is always bigger than k/n (since $n - 1$ is bigger than $n - 2$), so the bottom line is: always switch.

Exercise 9.1.1. We've seen that in any version of the Monty Hall problem it makes sense to switch doors. Can you give a conceptual (as opposed to numerical) explanation for why that would be?

9.2 CONDITIONAL PROBABILITY

All these Monty Hall problems illustrate the notion of *conditional probability*: we don't know if our initial guess was right or not, but we can calculate our chances in either case, and use this to determine the probability of winning.

To abstractify this, we need first some notation. In general, we'll write $P(A)$ for "the probability that an event A occurs." So, for example, if we're analyzing the Monty Hall problem with n doors and k cars, we could let A be the event "our initial guess is correct," and let B denote the event "our initial guess is wrong;" we have

$$P(A) = \frac{k}{n} \quad \text{and} \quad P(B) = \frac{n-k}{n}.$$

Note that in general, $P(A)$ will be a number between 0 and 1, with $P(A) = 1$ if A is a certainty, and $P(A) = 0$ if A can never occur. Also note that if exactly one of the two possibilities A and B must occur, but not both, then we must have

$$P(A) + P(B) = 1,$$

as in this case.

Now, suppose we're considering another event—in the Monty Hall example, say, winning the game, if we adopt the strategy of switching—which we'll denote by W. It may be, as it is in this case, that we don't know a priori the probability of W occurring, but that we do know it *if we assume that A occurs*. In these circumstances we write $P(W \text{ assuming } A)$ for the probability that W occurs, assuming that A is the case.

Again, in the Monty Hall example, this is the case: if we assume that our initial guess was correct, we calculated the probability of winning if we switch; we saw, in our new notation, that

$$P(W \text{ assuming } A) = \frac{k-1}{n-2}$$

and likewise

$$P(W \text{ assuming } B) = \frac{k}{n-2}.$$

One more bit of notation: if A and W are events, we'll write $P(W \text{ and } A)$ for the probability that *both* A and W occur. As a reality check, observe that

$$P(W \text{ and } A) = P(A) \cdot P(W \text{ assuming } A):$$

the probability of both A and W occurring is the probability that A will occur, times the probability that W will occur given that A is true.

Now, suppose we're in a situation where either A or B occurs, but not both. Numerically, this corresponds to the condition that $P(A) + P(B) = 1$. In that case, to say that W occurs is to say that either W and A both occur, or W and B both occur: in other words,

$$P(W) = P(W \text{ and } A) + P(W \text{ and } B).$$

Moreover, since $P(W \text{ and } A) = P(A) \cdot P(W \text{ assuming } A)$ and likewise for B, we can rewrite this as a general formula:

In a situation where either of two events A or B occurs, but not both, and W is a third event, whose outcome may depend on A or B:

$$P(W) = P(A) \cdot P(W \text{ assuming } A) + P(B) \cdot P(W \text{ assuming } B).$$

In words, assume that event A occurs the fraction $P(A)$ of the time, and that of the times A occurs, W will occur $P(W \text{ assuming } A)$ of the time; and likewise that event B occurs $P(B)$ of the time, and of these times W occurs $P(W \text{ assuming } B)$ of the time. Then the total fraction $P(W)$ of the time that W occurs is the likelihood $P(A) \cdot$

$P(W$ assuming $A)$ that both A and W occur; plus the likelihood $P(B) \cdot P(W$ assuming $B)$ that both B and W occur. This is exactly the calculation we made in figuring out the probability of winning at Monty Hall if we adopt the strategy of switching.

For example, if $P(A) = P(B) = 1/2$—that is, A and B occur equally often—the likelihood that we win is the average of $P(W$ assuming $A)$ and $P(W$ assuming $B)$, which makes sense. And as $P(A)$ increases and $P(B)$ decreases (since either one or the other must occur, the sum must be 1), we get a weighted average, with more weight on $P(W$ assuming $A)$; again, this makes sense.

In this setting, we call $P(W$ assuming $A)$ the *conditional probability* of winning, assuming that A occurs; likewise, $P(W$ assuming $B)$ is the conditional probability of winning given that B occurs.

Naturally enough, there's a more general version of this, in which exactly one of several events $A_1, \ldots, A_n$ must occur; we let $P(A_i)$ be the probability that A_i occurs. Suppose that if A_i occurs, our probability of winning is $P(W$ assuming $A_i)$. Then the chance $P(W)$ we'll win is

$$P(W) = P(A_1)P(W \text{ assuming } A_1) + \cdots + P(A_n)P(W \text{ assuming } A_n).$$

Finally, a note about typography: many sources will use the symbol $P(W|A)$ for $P(W$ assuming $A)$, and $P(W \cap A)$ for $P(W$ and $A)$. We won't use these symbols, but if you look something up in another source you're likely to run across them.

Problem 9.2.1. Gamblers Nathan and Carl are, for sheer lack of imagination, playing a simple game: they each roll a single die, and the higher one wins. If it's a tie, then to break the tie Nathan rolls a single die; if it comes up 1, 2, 3, or 4 Nathan wins, and if it comes up 5 or 6 Carl wins. How much of an edge does this give Nathan? In other words, what fraction of the time will he win?

Solution. There are three possible results from the initial roll: Nathan might win outright; Carl might win outright, or they could tie. We'll call these results A_N, A_C, and A_T respectively, and the first thing we need to do is to determine the likelihood of each of them.

This is straightforward. There are 36 possible outcomes of the initial roll by Nathan and Carl. In 6 of these we have a tie; the remaining 30 are split evenly between outcomes where Nathan wins and outcomes where Carl wins. Thus,

$$P(A_N) = \frac{15}{36}; \quad P(A_C) = \frac{15}{36}, \quad \text{and} \quad P(A_T) = \frac{6}{36}.$$

Next, what is the probability of Nathan winning, given each of these results from the first roll? Again, not difficult: if A_N occurs, Nathan wins outright; in other words (or symbols), if we denote by W the event that Nathan wins the game,

$$P(W \text{ assuming } A_N) = 1.$$

Likewise, if A_C occurs, Nathan has no chance of winning; that is,

$$P(W \text{ assuming } A_C) = 0.$$

Finally, if A_T occurs—the outcome of the first roll is a tie—then Nathan will win 4 times out of 6, so that

$$P(W \text{ assuming } A_T) = \frac{4}{6}.$$

Now we just have to add it all up: extrapolating from the formula on page 112,

$$P(W) = P(A_N)P(W \text{ assuming } A_N) + P(A_C)P(W \text{ assuming } A_C)$$
$$+ P(A_T)P(W \text{ assuming } A_T)$$

$$= \frac{15}{36} \cdot 1 + \frac{15}{36} \cdot 0 + \frac{6}{36} \cdot \frac{4}{6}$$
$$= \frac{19}{36}.$$

In other words, Nathan will win this game 19/36, or about 52.8%, of the time. □

Here's another gambling game that illustrates the notion of conditional probability:

Problem 9.2.2. Nathan and Carl have regressed to flipping coins. The game is as follows: Nathan picks a coin at random from a bag containing three coins, and flips it; if it's heads, Nathan wins, and if it's tails Carl wins. The kicker is, of the three coins in the bag, two are fair, meaning that they are equally likely to come up heads or tails, but one coin is rigged: it comes up tails 60%, or 3/5, of the time and heads only 2/5 of the time. The question is, what is the probability that Nathan wins?

Solution. Since we don't know which coin Nathan's picked, we don't know the probabilities governing his flip; but we know the probabilities in either case, and so we can apply our formula. To go through the logic: 2/3 of the time, Nathan will pick one of the fair coins, and of these times, he'll win half; 1/3 of the time he's pick the rigged coin, and of these times he'll win only 2/5. In other words, the total fraction of the time Nathan wins is 1/2 of 2/3 plus 2/5 of 1/3, or

$$\frac{1}{2} \cdot \frac{2}{3} + \frac{2}{5} \cdot \frac{1}{3} = \frac{14}{30}.$$

In symbols: if A is the event that Nathan picks a fair coin and B the event that he picks the rigged coin, we have

$$P(A) = \frac{2}{3} \quad \text{and} \quad P(B) = \frac{1}{3}.$$

Now, if W represents the result that Nathan wins, the problem tells us that

$$P(W \text{ assuming } A) = \frac{1}{2} \quad \text{and} \quad P(W \text{ assuming } B) = \frac{2}{5};$$

applying the formula of page 112, we have

$$P(W) = P(A) \cdot P(W \text{ assuming } A) + P(B) \cdot P(W \text{ assuming } B)$$
$$= \frac{2}{3} \cdot \frac{1}{2} + \frac{1}{3} \cdot \frac{2}{5}$$
$$= \frac{14}{30},$$

as before. □

Exercise 9.2.3. Suppose you have two coins in a bag, one fair coin and one trick coin that has heads on both sides. Without looking, you take one coin from the bag at random and flip it. What is the probability that it comes up heads?

Exercise 9.2.4. As in the last problem you have two coins in a bag, one fair coin and one trick coin that has heads on both sides. Without looking, you take one coin from the bag at random and flip it, and it comes up heads. What is the probability the coin you chose was the fair coin?

9.3   INDEPENDENCE

This would be a good time to talk about the concept of *independence* of events, a notion we've encountered before but now can express more precisely with our new notation.

Recall that when we started talking about probability, the first example we discussed was a sequence of coin flips. We observed that if we flipped n coins in succession, there were 2^n possible outcomes and, assuming that the coin is fair—that on any given roll, the coin is equally likely to come up heads and tails—*all 2^n outcomes are equally likely.* We expressed this by saying that "the coin has no memory;" in other words, the result of the first coin flip does not affect the second, and so on.

To express this in our new notation, suppose that H_1 is the result that the first coin flip comes up heads, and T_1 that it comes up tails; let H_2 be the statement that the second flip comes up heads, and T_2 that it's tails. When we say that the second coin flip is unaffected by the result of the first, we're saying that

$$P(H_2) = P(H_2 \text{ assuming } H_1) = P(H_2 \text{ assuming } T_1) = \frac{1}{2},$$

and likewise

$$P(T_2) = P(T_2 \text{ assuming } H_1) = P(T_2 \text{ assuming } T_1) = \frac{1}{2}.$$

In other words, the probability that the second coin comes up heads or tails is the same if we assume the first came up heads or if we assume the first came up tails.

In general, suppose we have an event with two possible outcomes, A and B, and then another event, with outcomes W and L. If the probability that the outcome of the second is W doesn't depend on the result of the first, that is, if

$$P(W) = P(W \text{ assuming } A) = P(W \text{ assuming } B),$$

then we say that the two events are *independent*. In this case, the probability that *both* A and W occur is just the product $P(A)P(W)$:

$$P(W \text{ and } A) = P(A) \cdot P(W \text{ assuming } A)$$
$$= P(A)P(W).$$

It's not really clear why this is a reasonable mathematical formulation of the intuitive notion of independence of events—for one thing, shouldn't it also depend on the relationship between $P(L)$ and $P(L$ assuming $A)$ and $P(L$ assuming $B)$? We will investigate this further in Exercise 9.3.4, but for now, for reference, we record this strange definition in another "black box":

> Given two events, the first of which has two possible outcomes A or B, and the second of which has two possible outcomes W or L, if
>
> $$P(W) = P(W \text{ assuming } A) = P(W \text{ assuming } B),$$
>
> then the two events are *independent*.

A sequence of coin flips is a good example of independent events. Dealing a sequence of cards from a deck, by way of contrast, is an example of a series of events that are not. For example, say that A represents the outcome that the first card you're dealt is a spade, B the outcome that it's one of the three other suits, and W the outcome that the second card is a spade. If we assume A—that is, we assume that the first card is a spade—then the remaining 51 cards in the deck consist of 12 spades and 39 non-spades; the chance that the second card will be another spade is thus

$$P(W \text{ assuming } A) = \frac{12}{51}.$$

On the other hand, if the first card is not a spade, then the remaining 51 cards consist of 13 spades and 38 non-spades, so that

$$P(W \text{ assuming } B) = \frac{13}{51}.$$

Here's a reality check: if we make no assumptions about the first card, the probability that the second card is a spade should be the probability that any card randomly chosen from a full deck of 52 is a spade, that is, 1 in 4; and indeed, since the probability $P(A)$ that the first card is a spade is 1 in 4, we have

$$P(W) = P(A) \cdot P(W \text{ assuming } A) + P(B) \cdot P(W \text{ assuming } B)$$

$$= \frac{1}{4} \cdot \frac{12}{51} + \frac{3}{4} \cdot \frac{13}{51}$$

$$= \frac{1 \cdot 12 + 3 \cdot 13}{4 \cdot 51}$$

$$= \frac{1}{4}.$$

For any pair of events A and W, if $P(W \text{ assuming } A) > P(W)$—that is, if W is more likely to occur given that A is the case than in general—we say that W *positively correlated* with A; if $P(W \text{ assuming } A) < P(W)$ (as in this last example), we say W is *negatively correlated* with A. For example, the last calculation says that, when you're dealt two cards from a standard deck, the probability that the second card is a spade is negatively correlated with the probability that the first card is a spade.

Problem 9.3.1. Consider again the multiple-door, multiple-car version of the Monty Hall game: that is, n doors and k cars, as in Section 9.1 above. Assuming we adopt the strategy of switching, is the probability of winning positively or negatively correlated with the probability that our initial guess was correct?

Solution. We've already worked out the relevant probabilities, so this is just a matter of making sure we know the meaning of "positively correlated" and "negatively correlated." It helps to set up notation first: we'll denote by A_r the event that our initial guess is right, and by A_w the event that our initial guess is wrong. (As we noted in Section 9.2, $P(A_r) = \frac{k}{n}$ and $P(A_w) = \frac{n-k}{n}$, though these won't be relevant to the current question.) As we saw in Section 9.1, if our initial guess is right and we switch, the probability that we win is $\frac{k-1}{n-2}$, while if our initial guess is wrong, then after switching we'll win $\frac{k}{n-2}$ of the time. In symbols, if we denote by W the event of our winning the car,

$$P(W \text{ assuming } A_r) = \frac{k-1}{n-2};$$

whereas we worked out in Section 9.1 that the overall probability of winning is

$$P(W) = \frac{k(n-1)}{n(n-2)}.$$

Now, this number is bigger than $P(W \text{ assuming } A_r) = \frac{k-1}{n-2}$ (check this yourself—or just read down to Problem 9.3.2), meaning that our chances of winning the car are worse if our initial guess is right. We say accordingly that the probability of winning the car is negatively correlated with the event that our initial guess was correct. $\square$

Problem 9.3.2. Just now, we said that $\frac{k(n-1)}{n(n-2)}$ was always bigger than $\frac{k-1}{n-2}$. How do we know this?

Solution. Always, when we want to compare fractions, we put them over a *common denominator* and multiply out: in this case, that means we write the fractions $\frac{k-1}{n-2}$ and $\frac{k(n-1)}{n(n-2)}$ as

$$\frac{k-1}{n-2} = \frac{k-1}{n-2} \cdot \frac{n}{n} = \frac{(k-1)n}{(n-2)n} = \frac{kn-n}{(n-2)n}$$

and

$$\frac{k(n-1)}{n(n-2)} = \frac{kn-k}{(n-2)n}.$$

Now that the denominators are equal, we just compare the numerators: since k is less than n, we see that $kn - k$ is less than $kn - n$ (we're subtracting a smaller quantity), and we conclude that

$$\frac{k(n-1)}{n(n-2)} > \frac{k-1}{n-2}. \qquad \square$$

Exercise 9.3.3. Suppose you're playing two-card poker: you're dealt two cards at random from a deck. We'll say you have a *mini-flush* if they're the same suit, and a *mini-straight* if their denominations are in sequence. (Note that both A2 and AK count as mini-straights.)

1. First of all, what is the probability of getting a mini-flush?
2. What is the probability of getting a mini-straight?
3. What is the probability that you have a mini-straight, assuming you have a mini-flush?
4. What is the probability that you have a mini-flush, assuming you have a mini-straight?

5. Are the events getting a mini-flush and getting a mini-straight independent, positively correlated or negatively correlated? Justify whichever answer you come up with.

We leave you with two conceptual questions that we don't yet have the tools to answer mathematically. We'll see the answer to both of them in Section 9.5, following a discussion of Bayes' theorem, but wanted to give you the opportunity to reason about them informally now.

Exercise 9.3.4. Consider two events, the first of which has two possible outcomes A and B, and the second of which has two possible outcomes W and L.

1. If W is independent the outcome of the first event, is L also independent of the outcome of the first event?
2. If $P(W) = P(W \text{ assuming } A) = P(W \text{ assuming } B)$ does this necessarily mean that $P(L) = P(L \text{ assuming } A) = P(L \text{ assuming } B)$?
3. What do these questions have to do with each other?

Here's another question along similar lines:

Exercise 9.3.5. Is the relationship of correlation a symmetrical one? In other words, if W is positively correlated with A, is it necessarily the case that A is positively correlated with W?

9.4   AN ELECTION

For another example of a situation where conditional probability arises, suppose a student body election is coming up. There are two candidates, Tracy and Paul. Polling has revealed a major difference in the support of the two candidates among two classes of voters: the left-handed and the right-handed. Possibly because of Tracy's forceful advocacy of left-handed desks in classrooms, among left-handed voters 75% support Tracy and only 25% support Paul. Right-handed voters are not convinced, however: among these voters, 60% support Paul and only 40% support Tracy. Say that lefties constitute 20% of the voting population. Who wins the election? To phrase the question differently (but equivalently), if you ask a random voter, what is the probability that they'll be voting for Tracy?

We can break it down much as we did in the Monty Hall problem. Lefties represent one-fifth of the population, and among them three-quarters are voting for Tracy; thus the fraction of the population who are left-handed *and* supporters of Tracy is

$$\frac{1}{5} \cdot \frac{3}{4} = \frac{3}{20} = \frac{15}{100},$$

or 15%. Likewise, the proportion of left-handed supporters of Paul is

$$\frac{1}{5} \cdot \frac{1}{4} = \frac{1}{20} = \frac{5}{100},$$

or 5%. We can do the remaining cases similarly, and put the information in a table:

	Tracy	Paul
lefties	$\frac{1}{5} \cdot \frac{3}{4} = \frac{15}{100}$	$\frac{1}{5} \cdot \frac{1}{4} = \frac{5}{100}$
righties	$\frac{4}{5} \cdot \frac{2}{5} = \frac{32}{100}$	$\frac{4}{5} \cdot \frac{3}{4} = \frac{48}{100}$
total	$\frac{47}{100}$	$\frac{53}{100}$

In other words, out of 100 randomly selected voters, we'd expect 32 to be right-handed Tracy supporters, 48 to be right-handed Paul voters, 15 to be lefties voting for Tracy and 5 lefties supporting Paul. Adding it up, we see that Tracy has 47% of the vote, and Paul 53%. Tracy loses, showing the dangers of appealing to too narrow a base of voters.

Again, this illustrates the same principle we saw in the Monty Hall problem. To say it (at least partially) in English, we can put it this way: since a given voter is either right-handed or left-handed,

> The probability that a random voter X is voting for Tracy

is

> the probability that X is left-handed and is voting for Tracy

plus

> the probability that X is right-handed and is voting for Tracy;

which is

> the probability that X is left-handed times the probability that X is voting for Tracy, *given that she is left-handed*

plus

> the probability that X is right-handed times the probability that X is voting for Tracy, *given that she is right-handed*;

which is

$$\frac{1}{5} \cdot \frac{3}{4} + \frac{4}{5} \cdot \frac{1}{4},$$

or, clearing denominators and multiplying out,

$$\frac{47}{100}.$$

We can express this in terms of the notation we introduced in Section 9.2. Let R be the event that a randomly selected student is right-handed, and L the condition that she's left-handed, so that according to the setup

$$P(R) = \frac{4}{5} \quad \text{and} \quad P(L) = \frac{1}{5}.$$

Likewise, let T be the statement that our randomly selected student is a Tracy supporter, and P the statement that she's voting for Paul. The results of our polling, as stated in the problem, are that 75% of lefties favor Tracy, or in other words

$$P(T \text{ assuming } L) = \frac{3}{4} \quad \text{and} \quad P(P \text{ assuming } L) = \frac{1}{4}.$$

Similarly, the problem states that only 40% of righties favor Tracy, which we write as

$$P(T \text{ assuming } R) = \frac{2}{5} \quad \text{and} \quad P(P \text{ assuming } R) = \frac{3}{5}.$$

In these terms, our calculation of the probability $P(T)$ that a randomly selected student is voting for Tracy can be expressed as an application of the general formula from Section 9.2 above:

$$P(T) = P(L) \cdot P(T \text{ assuming } L) + P(R) \cdot P(T \text{ assuming } R)$$
$$= \frac{1}{5} \cdot \frac{3}{4} + \frac{4}{5} \cdot \frac{1}{4}$$
$$= \frac{47}{100}.$$

Problem 9.4.1. Suppose that lefties make up 40% of the population rather than 20%. Assuming Tracy and Paul have the same percentage support as before among each group, who wins the election now?

Solution. As in the original version, we start by working out the fraction of all voters who fall into each of the four types: left-handed Tracy supporters, left-handed Paul voters, right-handed for Tracy, and right-handed Paul supporters. We do this is the same way: for example, we know that 30% of the population is left-handed, and of these three-quarters are Tracy supporters; thus lefties for Tracy comprise a total of

$$\frac{3}{4} \cdot \frac{40}{100} = \frac{30}{100}$$

of the population, and the fraction of the population who are left-handed Paul voters is

$$\frac{1}{4} \cdot \frac{40}{100} = \frac{10}{100}.$$

Similarly, the 60% of the population who are right-handed break down into

$$\frac{2}{5} \cdot \frac{60}{100} = \frac{24}{100}$$

who support Tracy, and

$$\frac{3}{5} \cdot \frac{60}{100} = \frac{36}{100}$$

who support Paul. Putting this all in a table as before, we have

	Tracy	Paul
lefties	$\frac{30}{100}$	$\frac{10}{100}$
righties	$\frac{24}{100}$	$\frac{36}{100}$
total	$\frac{54}{100}$	$\frac{46}{100}$

and adding up the columns we see that Tracy now has 54% of the vote, and Paul 46%—showing that pandering does work, as long as the segment of the population you're pandering to is large enough. $\square$

Exercise 9.4.2. In the election described above, decide whether each pair of events listed below are *positively correlated*, *negatively correlated*, or *independent*.

1. A voter being left-handed and preferring Tracy.
2. A voter being right-handed and preferring Tracy.
3. A voter being left-handed and preferring Paul.
4. A voter being right-handed and preferring Paul.

9.5 BAYES' THEOREM

We want to continue a little further with the election of the last example (as originally described at the beginning of Section 9.4), and tease out a relationship among the probabilities involved. To start with, two questions:

Problem 9.5.1. Say you sit down at the cafeteria and, observing that the person opposite you holds her spoon in her left hand, deduce that she's left-handed. What is the probability that she supports Tracy?

Solution. Assuming, as we shall, that only lefties would eat with a spoon in their left hand, to answer this question, we need only reread the statement of the problem that 75% of lefties support Tracy. But now:

Problem 9.5.2. Say you sit down at the cafeteria and see that the person opposite you is wearing a large button that says, "I $\heartsuit$ Tracy." What is the probability that she's left-handed?

Solution. OK, this one requires us to think a little. But before we do, we want to point out that *the two questions we've just asked are not the same problem*! This is absolutely one of the most common mistakes people make in estimating probabilities: assuming that the answers to Problems 9.5.1 and 9.5.2 are the same.[3] It's a completely reasonable mistake, in that it makes sense in a sort of qualitative way: other things being equal, the larger the overlap of the class of lefties and the class of Tracy supporters, the more apt

---

[3] As we were writing this in 2012, we thought we'd look at that week's papers and see if we could find an example. It took exactly five minutes: a columnist online, bemoaning the Dallas Cowboys' 1-4 record, wrote that "Only 4% of all playoff teams have started the season 1-4," and then continued with, "That means the Cowboys have only a 1 in 25 chance of making the playoffs." No, no, no! We'll discuss this in more detail, and see how to fix the columnist's error, in Problem 9.5.3 below.

a member of one group is to be a member of the other. *But it's not true.* We'll see lots of examples of this, starting with this one, and we'll also see how to fix it.

Back to our problem. As we've worked out, among 100 students we'd expect 47 to support Tracy. How many of these are left-handed? We already worked that out—it's in the table above—but we'll remind you: of the 100, we expect 20 to be lefties, and three-quarters of these, or 15, to be Tracy supporters. So the probability of a randomly chosen Tracy supporter being left-handed is

$$\frac{15}{47} \sim 0.32,$$

or less than 1 in 3—much less than the answer to Question 1!

In fact, the relations among these probabilities can all be read off of the table we made above, and which we'll reproduce here:

	Tracy	**Paul**	**total**
lefties	$\frac{1}{5} \cdot \frac{3}{4} = \frac{15}{100}$	$\frac{1}{5} \cdot \frac{1}{4} = \frac{5}{100}$	$\frac{20}{100}$
righties	$\frac{4}{5} \cdot \frac{2}{5} = \frac{32}{100}$	$\frac{4}{5} \cdot \frac{3}{4} = \frac{48}{100}$	$\frac{80}{100}$
total	$\frac{47}{100}$	$\frac{53}{100}$	

Now we'll drop all the $\frac{1}{100}$s and just give the expected number of each type among 100 random students:

	Tracy	**Paul**	**total**
lefties	15	5	20
righties	32	48	80
total	47	53	

Focus on the 2 × 2 box in the middle of the table. Question 1 is asking, "Of all the people in the first *row*, how many belong to the first box?" Question 2 is asking, "Of all the people in the first *column*, how many belong to the first box?" The point is, the number of people in the first box, 15, is the same in both; we're just dividing by different numbers: in the first, we're dividing by the number 20 people in the first row; in the second, the number 47 of people in the first column. The conclusion is the answers to Questions 1 and 2 differ by the ratio of these two numbers: in numbers,

$$\frac{15}{47} = \frac{15}{20} \cdot \frac{20}{47},$$

where the last number represents the ratio of the likelihood that a random student is a Tracy supporter to the likelihood that she's left-handed.

This relationship between the answers to Questions 1 and 2 is a special case of *Bayes' theorem*, a central observation in probability.

The basic situation is that we have a population—the outcomes of a series of dice rolls or coin flips, or the selection of a random student—that is divided in two ways: everyone is either in Camp A or Camp B, and everyone is either in Camp M or Camp N. Suppose that members of Camp A comprise the fraction $P(A)$ of the population—in other words, let $P(A)$ be the probability that a randomly chosen member of our population belongs to Camp A—and likewise let $P(M)$ be the fraction of the population in Camp M.

Suppose now we want to know the likelihood that a randomly chosen outcome belong to *both* Camp A and Camp M—that is, we want to calculate the probability $P(A \text{ and } M)$. We have two ways of doing this: first, we can write

$$P(A \text{ and } M) = P(M) \cdot P(A \text{ assuming } M);$$

but since $P(A \text{ and } M)$ is the same thing as $P(M \text{ and } A)$ we can also write this as

$$P(A \text{ and } M) = P(A) \cdot P(M \text{ assuming } A).$$

Equating these two, we have *Bayes' theorem*:

Given one event with outcomes A and B and a second event with outcomes M and N:

$$P(A \text{ and } M) = P(A) \cdot P(M \text{ assuming } A) = P(M) \cdot P(A \text{ assuming } M).$$

Consequently:

$$P(M \text{ assuming } A) = P(A \text{ assuming } M) \cdot \frac{P(M)}{P(A)}.$$

Or in words: the probability that a randomly chosen member of Camp A belongs to Camp M is equal to the probability that a randomly chosen member of Camp M belongs to Camp A, times $P(M)/P(A)$.

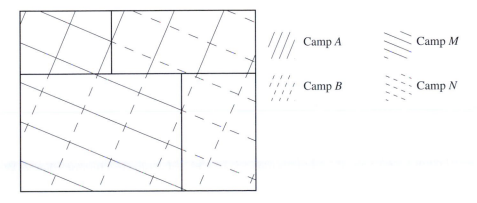

Problem 9.5.3. Let's bring back the Dallas Cowboys and their 1-4 record, mentioned in the last footnote, and see how we can fix the mistake the columnist made. To start

with, the population is the collection of football teams, where a "team" means a given franchise *in a given year*—the '97 Dolphins, say, or the '04 Pats. Camp M will be the teams that make the playoffs; Camp N be will be the teams who don't. Similarly, Camp A will be all teams that start the season 1-4; and Camp B will be the teams that don't. The columnist starts out with the assertion that "of all teams that have gone to the playoffs, only 4% have started the season 1-4"—in other words,

$$P(A \text{ assuming } M) = \frac{4}{100} = \frac{1}{25}.$$

Let's assume this is correct. The question now is, what is the probability that a team that's started the season 1-4 will go to the playoffs—that is, what is $P(M \text{ assuming } A)$? Again, *that's not the same* as $P(A \text{ assuming } M)$, as the columnist seems to think; but Bayes will allow us to recover.

The information we'll need in order to apply Bayes' theorem is $P(M)$ and $P(A)$—the probability that a random team will go to the playoffs, and the probability that a random team will start the season 1-4—and we can estimate these. To begin with, each year (under the current playoff system), 12 teams out of the 32 NFL franchises go to the playoffs, so the probability that a random team is going to the playoffs is

$$P(M) = \frac{12}{32} = \frac{3}{8} = 0.375.$$

We don't know what fraction of all teams start the season 1-4, so let's wing it. If the outcome of each game was decided by a coin flip, the probability of a team winning exactly one of its first five games would be

$$P(A) = \frac{\binom{5}{1}}{2^5} = \frac{5}{32} \sim 0.156.$$

The actual fraction is probably slightly higher—games are not decided by coin flips; there are good teams and bad teams, and a team that's already gone 1-3 is apt to be a bad team and thus more likely to go to 1-4 than to 2-3. But we'll use this as a first guess; it's probably not too far off.

Given all this, Bayes provides the antidote to our columnist's mistake: the probability that a team that has gone 1-4 (that is, is a member of Camp A) will go to the playoffs (that is, be a member of Camp M), according to Bayes' theorem, equal to the likelihood that a team going to the playoffs (member of Camp M) has gone 1-4 (Camp A), times the ratio $P(M)/P(A) = 0.375/0.156 = 2.4$. Assuming the first statement of the columnist—that the likelihood of a playoff team having started 1-4 is .04—is accurate, then the likelihood of a 1-4 team going to the playoffs must be

$$.04 \times 2.4 = .096,$$

or nearly one in 10! The Cowboys are not so bad off after all.[4]

We'll end this section with a short note on correlations and independence. We said in Section 9.3 that an event M was *positively correlated* with an event A if

$$P(M \text{ assuming } A) > P(M)$$

---

[4] Actually, they were, but that's not the point.

that is, M is more likely to occur if A is the case than otherwise. Though it wasn't necessarily obvious at the time, one immediate consequence of Bayes' theorem is that *this is a symmetrical relationship*: that is, M is more likely to occur when A is the case exactly if A is more likely to occur when M is the case. To see this, simply divide both sides of Bayes' theorem by the product $P(A)P(M)$ to express it in the form

$$\frac{P(A \text{ assuming } M)}{P(A)} = \frac{P(M \text{ assuming } A)}{P(M)}.$$

To say that M is positively correlated with A is exactly to say that the right-hand side of this equation is bigger than 1. If that's the case, so is the left-hand side, which is to say A is positively correlated with M.

   This is probably one of the reasons people so often mistakenly equate the two quantities $P(A \text{ assuming } M)$ and $P(M \text{ assuming } A)$: they confuse the qualitative (and true) statement that M is positively correlated with A if and only if A is positively correlated with M, with the quantitative (and false, false, false!) statement that $P(A \text{ assuming } M)$ is the same as $P(M \text{ assuming } A)$.

Exercise 9.5.4. We're playing three-card poker—that is, we're dealt three cards in a row at random from a standard deck.

1. What is the probability of getting a pair?
2. Suppose your first two cards are of different denominations. Now what's the probability of getting a pair?
3. Suppose you were dealt a pair in three cards. What are the odds that you had it on your first two cards—that is, that your first two cards were of the same denomination?

Exercise 9.5.5. You own a very nice umbrella, which keeps you totally dry, but it's also rather heavy, so you don't like to carry it every day. You also have bad luck with the weather. Five days out of the seven days in a week, you carry your umbrella, but the other two days you decide to risk it and leave the umbrella at home. When you're carrying the umbrella, it rains 1/3 of the time, but when you are not carrying the umbrella it rains 5/6 of the time.
It's raining today. What is the probability that you are carrying your umbrella?

Exercise 9.5.6. Go back and solve Exercises 9.3.4 and 9.3.5.

9.6   THE ZOMBIE APOCALYPSE

Many of the adults in Jamestown work in the quarry, so at any given time 40% of them are covered in dirt. Likewise, 90% of the zombies in Jamestown are covered in dirt, though the remaining 10% have managed to clean themselves up somehow. Overall, 20% of the population of Jamestown are zombies. Here's the question:

Problem 9.6.1. Suppose you see a figure in the distance. You are not close enough to discern whether they are human or a zombie, but you can see that they are covered in dirt. What is the probability that a zombie is approaching you? Should you run?

Solution. As always, we start by giving names to the various possibilities: we'll denote by H the statement that a figure in the distance is a human and by Z the statement that the figure is a zombie. Similarly, we'll let D be the statement that a figure in the distance is covered in dirt and C the statement that the figure in the distance is not covered in dirt (is clean).

The problem tells us the probabilities

$$P(D \text{ assuming } H) = 40\% \quad \text{and} \quad P(D \text{ assuming } Z) = 90\%.$$

The catch is that instead we'd like to compute $P(Z \text{ assuming } D)$. To do so, we'll have to use Bayes' theorem.

We know that 20% of the population of Jamestown are zombies, so we can use this information to compute the probability that any individual in the town is *both* a zombie and covered in dirt:

$$P(D \text{ and } Z) = P(Z) \cdot P(D \text{ assuming } Z) = \frac{1}{5} \cdot \frac{9}{10} = \frac{9}{50}.$$

Similarly, we can compute the probability that any individual in the town is *both* a human and covered in dirt:

$$P(D \text{ and } H) = P(H) \cdot P(D \text{ assuming } H) = \frac{4}{5} \cdot \frac{2}{5} = \frac{8}{25}.$$

Now using our conditional probability formula from Section 9.2, we can compute the overall probability that any resident of Jamestown is covered in dirt:

$$P(D) = P(D \text{ and } Z) + P(D \text{ and } H) = \frac{9}{50} + \frac{8}{25} = \frac{1}{2}.$$

Finally, we apply Bayes' theorem again in its equivalent formulation to compute the likelihood that an approaching figure in the distance who is covered in dirt is a zombie:

$$P(Z \text{ assuming } D) \;=\; P(D \text{ assuming } Z) \cdot \frac{P(Z)}{P(D)} = \frac{9}{10} \cdot \frac{\frac{1}{5}}{\frac{1}{2}} = \frac{9}{25},$$

which is about 36%, so your chances of survival aren't too bad on the whole. □

Problem 9.6.2. Suppose that among households with an annual income over \$100,000, 75% own SUVs, while only 20% of households with an annual income under \$100,000 do. Suppose also that households making over \$100,000 make up one-fifth of all households in the country. What are the odds that an SUV driver's household income is over \$100,000?

Solution. As always, we start by figuring out the fraction of households in each of the four classes: rich folks with SUVs, rich folks without, and so on. According to the problem, the first of these comprise three-quarters of 20%, or 15%, of all households; rich folks without SUVs make up one-quarter of 20%, or 5%, of all households. Meanwhile, of the 80% of households with income under \$100,000, one-fifth (that is, 16% of all households) have SUVs, while the other four-fifths (64% of all households) don't. Here are the percentages, in the by-now-familiar table

	SUVs	no SUV	total
more than $100K	15%	5%	20%
less than $100K	16%	64%	80%
total	31%	69%	

To answer the question, we look in the left column, under "SUVs:" we see that a total of 31% of households own SUVs, consisting of the 15% of all households making over $100K and owning a SUV, and the 16% of all households making under $100K but owning an SUV anyway. The odds that the guy in the Lincoln Navigator who just cut you off is rich is thus 15/31, or just under half. Thus, even though the probability of a rich person owning an SUV is high (75%, at least in this problem) the probability of an SUV driver being rich is less than half.   □

Now you try one:

Exercise 9.6.3. Among US households with incomes in the top 1%, 15% of families own a cottage in New Zealand (presumably, so that they may survive the zombie apocalypse). Among the other 99% of US households, 2% of families own a cottage in New Zealand. (In case it wasn't clear by now, these numbers are entirely made up.) What is the likelihood that an American owner of a cottage in New Zealand has an income in the top 1%?

9.7   FINALLY, TEXAS HOLD 'EM

Frankly, I think what dates us more than anything else—more even than the lame word problems—is the version of poker we describe. Draw poker—the only version to which our calculations up to now are relevant—is long gone, replaced first by stud and then by Texas Hold 'Em. Happily, Texas Hold 'Em provides some useful examples of conditional probability. We'll discuss these varieties of poker in this section, after briefly revisiting draw poker to warm ourselves up.

Problem 9.7.1. Say we're playing three-card poker.

1. What are the odds of getting dealt a pair (that is, two cards of the same denomination)?
2. What are the odds of getting a pair if your first two cards are of different denominations?
3. Assuming you were dealt a pair in three cards, what are the odds that you had it on your first two cards—that is, that your first two cards were of the same denomination?

Solution. The first part is a standard problem, one that we could have asked a while back. For the purpose of this problem, since the order in which we receive the cards does matter, we'll take a hand to be a sequence—rather than a collection—of three cards. There are thus $52 \cdot 51 \cdot 50 = 132{,}600$ possible hands, all equally likely. The number with *no* pair is easy to calculate as well: we have 52 choices for the first card, then 48 for the second and 44 for the third, for a total of $52 \cdot 48 \cdot 44 = 109{,}824$ hands with no pair. The number of hands that do have a pair is thus $132{,}600 - 109{,}824 = 22{,}776$, and the probability of getting a pair is correspondingly

$$\frac{22{,}776}{132{,}600} \sim 0.17.$$

The second part is easier: if our first two cards are of two different denominations, to get a pair our third card has to be one of the six cards remaining in the deck that match one of our first two. Since there are a total of 50 cards remaining, our chances are

$$\frac{6}{50} \sim 0.12.$$

In other words, as we'd expect our chances of getting a pair are a good bit less if we haven't paired up on the first two cards.

Finally, the problem asks us: *among all hands that do have a pair, what fraction have a pair on the first two cards?* Again, we can calculate this in straightforward fashion: to get a hand with a pair on the first two cards, we have 52 choices for the first card, then 3 choices for the second, and finally the third card can be any of the remaining 50. The number of such hands is thus $52 \cdot 3 \cdot 50 = 7{,}800$. Thus, if we assume that our hand has a pair, the probability that we had it on the first two cards is

$$\frac{7{,}800}{22{,}776} \sim 0.34. \qquad \square$$

In many versions of poker, each player has access to seven cards rather than five; the player is allowed to choose the best five-card hand from the seven cards available. Examples of this type of poker include seven-card stud and Texas Hold 'Em; we'll be analyzing the latter, and so we should start by calculating at least some of the relevant odds. You'd expect the average hand to be better; we'll see in a moment by how much.

For example: what is the likelihood of getting a flush in seven-card poker? Meaning, of all $\binom{52}{7}$ possible collections of seven cards chosen from the standard deck of 52, how many include five cards of the same suit? This is something we know how to calculate, though it's a little more complicated than the five-card version.

To begin with, if we're going to specify a hand with five cards of one suit, we have to start by picking a suit to have five cards in; that's a choice of four. Say it's spades. Among the seven cards in our hand we could have five spades, six spades or seven spades, and we'll have to count the number of hands of each type.

- The number of hands with seven spades is the number of collections of seven cards chosen from the 13 spades in the deck; that's

$$\binom{13}{7} = 1{,}716.$$

- To specify a hand with exactly six spades, we have to specify a collection of six of the 13 spades, then choose one of the 39 non-spades in the deck; the number of such hands is thus

$$\binom{13}{6} \times 39 = 66{,}924.$$

- Similarly, to specify a hand with exactly five spades, we have to specify a collection of five of the 13 spades, then choose a collection of two of the 39 non-spades; the number of such hands is thus

$$\binom{13}{5} \times \binom{39}{2} = 1{,}287 \times 741 = 953{,}667.$$

The number of hands with at least five spades is therefore

$$953{,}667 + 66{,}924 + 1{,}716 = 1{,}022{,}307;$$

the number with at least five of some suit is 4 times that, or 4,089,228. And since the number of all possible collections of seven cards is

$$\binom{52}{7} = 133{,}784{,}560,$$

the odds of getting a flush are

$$\frac{4{,}089{,}228}{133{,}784{,}560} \sim 0.0306;$$

roughly three times in a hundred deals, in other words. As expected, flushes are far more frequent in seven-card poker than in five, where they occur only about one in 500 times.

Finally, Texas Hold 'Em works as follows: each player is dealt two cards face down; only that player knows what they are. Five cards are dealt face down in the middle; no one knows what they are. At the end, players will make the best five-card hand possible out of the two cards they hold, plus the five cards in the middle (all players can use these).

Here's how things proceed: there's a round of betting, then three of the five cards in the middle are turned up. Another round of betting; another card; another round of betting; another card; another round of betting, and finally the remaining players reveal their own cards and the winner is determined. We're mentioning all the betting not because we want to get into the mechanics of betting poker—life is way too short— but because at each of these stages we'll be required to assess our hand's potential, and calculate the odds of our getting three of a kind, or a flush or whatever, *based on incomplete information*. In other words, we'll know some of the cards in our hand, but not all. How does this extra information affect the odds?

To illustrate, suppose we sit down at the table and the cards are dealt. Before we look at our own cards—when we know nothing at all about our cards, or the ones in the middle of the table—the odds that we'll wind up with a flush are, as we just calculated, around one in 33. Now, suppose we look at our two cards, and see that they are of the same suit—hearts, say. What are the odds now of a flush? What would the odds of a flush be if we looked down and saw two cards of different suits?

Let's do the case of two hearts first. Basically, the cards in the middle of the table are a collection of five cards, chosen at random from among the 50 cards in the deck other than the two you're looking at. The question is, out of the $\binom{50}{5} = 2{,}118{,}760$ possible such collections, how many will produce a flush when combined with the two cards we hold?

There are two ways this could happen: the middle cards could include three or more hearts, or all five could belong to one of the other three suits. (The last possibility is highly unlikely, as we'll see, and in any case should be treated differently: in that case everybody at the table will have a flush.) For the former, we ask three questions:

- How many collections of five cards with exactly three hearts are there? Well, we have to specify a collection of three hearts from among the 11 remaining hearts in the deck, for $\binom{11}{3} = 165$ choices. Then we have to specify two cards from the 39 non-hearts, for $\binom{39}{2} = 741$ choices; there are thus a total of

$$\binom{11}{3}\binom{39}{2} = 165 \times 741 = 122{,}265$$

such possibilities for the five cards in the middle.

- How many collections of five cards with exactly four hearts are there? This is similar: we choose four hearts from among 11, then one card from among 39, for a total of

$$\binom{11}{4}\binom{39}{1} = 330 \times 39 = 12{,}870$$

such collections

- Finally, how many collections of five cards are there consisting entirely of hearts? This is is the easiest: it's just

$$\binom{11}{5} = 462.$$

Finally, it's possible that the cards in the middle will consist entirely of cards of one of the other three suits besides hearts, which can occur in

$$3 \times \binom{13}{5} = 3 \times 1{,}287 = 3{,}861$$

ways. Altogether, then, there are

$$122{,}265 + 12{,}870 + 462 + 3{,}861 = 139{,}458$$

ways in which the five cards in the center could give you a flush, out of the $\binom{50}{5} = 2{,}118{,}760$ possible collections; the odds that you'll get a flush are correspondingly

$$\frac{139{,}458}{2{,}118{,}760} \sim 0.06582,$$

or about 1 in 15. In other words, having two cards of the same suit a little more than doubles your chances of winding up with a flush.

OK, then; what if you look down and see that your two cards belong to different suits—say, you have one heart and one spade. As you might expect, your chances of getting a flush are now a good bit lower; let's see by how much.

Again, we just have to count how many collections of five cards, chosen from the remaining 50 cards in the deck, will combine with our two to make up a flush. As before, there are several ways they could do this: the cards in the middle could include

four or more hearts or spades; or they could be all clubs, or all diamonds. The number
with exactly four hearts is

$$\binom{12}{4}\binom{38}{1} = 495 \times 38 = 18{,}810$$

(remember there are 12 hearts, and 38 non-hearts, in the remaining 50 cards of the
deck); the number with five hearts is

$$\binom{12}{5} = 792,$$

so there are

$$18{,}810 + 792 = 19{,}602$$

ways of getting a heart flush, and the same number of ways of getting a flush in spades.
As for the possibility of all five cards in the middle being clubs, the number of such
collections is

$$\binom{13}{5} = 1{,}287$$

(as we worked out above), and similarly for diamonds. Altogether, then, the number of
collections of five cards that give us a flush is

$$2 \times 19{,}602 \, + \, 2 \times 1{,}287 = 41{,}778,$$

and the odds that we'll wind up with a flush are correspondingly

$$\frac{41{,}778}{2{,}118{,}760} \sim 0.01972,$$

or about 1 in 50.

Exercise 9.7.2. In the game of Texas Hold 'Em, two cards are dealt to you specifically
and then five cards are dealt face up on the table and available to all players. You may
choose the best five cards out of these seven to make a poker hand. Suppose the two
cards dealt to you are of different denominations. What is the probability you'll be able
to make four of a kind out of all seven cards available to you?

Exercise 9.7.3. In the game of Texas Hold 'Em, suppose there is a full house dealt on
the table. What is the probability that the two cards in your hand give you a better
five-card hand than the table has been dealt?

10 Life's like that: unfair coins and loaded dice

Say you're at a local pub, quaffing a few refreshing beverages, and the guy next to you proposes a game. A coin is to be flipped repeatedly; each time it comes up heads you pay him $1, and each time it comes up tails he pays you the same. You accept the proposal (perhaps you've quaffed a few too many refreshing beverages), but gradually become suspicious: of the first 10 coin flips, 7 come up heads. What are the chances that the coin is rigged?

This is a trick question: as we'll see when we discuss it, you simply don't have enough information to answer it. But it's a starting point for our next discussion, which is about the probabilities associated to reiterated events (coin flips, dice rolls and the like) when the odds on each event are skewed.

10.1 UNFAIR COINS

In our initial discussion of the probabilities associated to series of coin flips, we made two fundamental assumptions:

- the coin flips were independent events (the outcome of each flip has the same probability of coming up heads or tails, regardless of the outcomes of the other flips); and
- the coin was *fair*: that is, the probability of heads on a given flip was exactly 1/2.

Given these, we argued, there were 2^n possible outcomes of a series of n coin flips, and *they were all equally likely*, meaning each will occur $1/2^n$ of the time. Thus, for example, the probability of getting exactly k heads in n flips would be

$$\frac{\binom{n}{k}}{2^n},$$

as we discovered in the formula on page 48.

We're now going to alter the second of our basic assumptions, and ask: what if the coin is unfair? To be concrete, suppose that each time it's flipped, the probability that it comes up heads is p, and the probability that it comes up tails is $q = 1 - p$, but it's not necessarily the case that $p = q = 1/2$. Can we calculate the probability of getting exactly k heads in n flips now?

The answer is "yes," but we before we do this let's point out that the use of the pejorative term "unfair" is unfortunate, as is the whole association with coin flipping. In fact, what we're going to work out now applies much more broadly, to any independent sequence of iterations of an event with exactly two possible outcomes. For this reason, for our first example we'll use lottery tickets, in a simple lottery where there only two kinds of tickets, winning and losing.

Samuel Pepys once asked Isaac Newton the following question (or something like it). If the probability of a lottery ticket winning a prize are 1 in 6, which is most likely: to win at least one prize after buying six tickets; to win at least two prizes after buying 12 tickets; or to win at least three prizes after buying 18 tickets?

Let's answer Pepys' question by calculating the probabilities in each of his three cases, starting with buying six tickets. In fact, we'll do a little more than we have to here: we'll say what the likelihood is that, among our six tickets, there are any given number of winners.

Now, if we buy six tickets, we can view the outcome as a sequence of six letters, consisting of Ws (for winning tickets) and Ls (for losing ones). There are thus 2^6 possible outcomes, as with a sequence of six coin flips. But, unlike the coin flips, they are not all equally likely! If we assume the outcomes of the individual tickets are independent events (which they are, in effect, if the lottery is large enough), then the chance of all six being winners is

$$P(WWWWWW) = \left(\frac{1}{6}\right)^6.$$

By contrast, the likelihood of six being losing tickets is

$$P(LLLLLL) = \left(\frac{5}{6}\right)^6.$$

For a less simple example, consider the outcome represented by $LLWLWL$—in other words, the first two tickets are losers, the third wins, and so on. Now, the probability of the first ticket being an L is $5/6$, and ditto for the second; the probability of the third being a winner is $1/6$, and likewise for the remaining three. All in all, the probability of this exact outcome is

$$\frac{5}{6} \cdot \frac{5}{6} \cdot \frac{1}{6} \cdot \frac{5}{6} \cdot \frac{1}{6} \cdot \frac{5}{6} = \left(\frac{1}{6}\right)^2 \left(\frac{5}{6}\right)^4,$$

the exponents 2 and 4 corresponding to the number of Ws and Ls in the word, respectively. The probability of any word consisting of k Ws and $6-k$ Ls occurring is similarly

$$P(W^k L^{6-k}) = \left(\frac{1}{6}\right)^k \left(\frac{5}{6}\right)^{6-k},$$

where "$W^k L^{6-k}$" stands for "any word consisting of k Ws and $6-k$ Ls;" and (might as well go whole hog) more generally still, the probability of a specific sequence consisting of a Ws and b Ls occurring is

$$P(W^a L^b) = \left(\frac{1}{6}\right)^a \left(\frac{5}{6}\right)^b.$$

Now, what is the probability of getting no winners at all? We already worked that out: if we have no winners, the outcome must be LLLLLL, which occurs with probability

$$P(0 \text{ winners in } 6) = P(L^6) = \left(\frac{5}{6}\right)^6 \sim 0.335.$$

What about the likelihood of getting exactly 1 winner? Well, there are $\binom{6}{1}$ outcomes consisting of 5 *L*s and one *W*, and each occurs with probability $(1/6)(5/6)^5$, so the probability of exactly one winner is

$$P(1 \text{ winner in } 6) = \binom{6}{1}\left(\frac{1}{6}\right)\left(\frac{5}{6}\right)^5 \sim 0.402.$$

Note that this is slightly more likely than getting no winners at all: while the probability of a given outcome involving 1 *W* and 5 *L*s is one-fifth the probability of all losers, there are now six possible outcomes rather than one.

Next, the probability a given sequence with two winners is $(1/6)^2(5/6)^4$, and there are $\binom{6}{2} = 15$ such sequences, so

$$P(2 \text{ winners in } 6) = \binom{6}{2}\left(\frac{1}{6}\right)^2\left(\frac{5}{6}\right)^4 \sim 0.201.$$

Again, it makes sense that this is half the probability of getting one winner: the probability of a given sequence with 2 *W*s occurring are one-fifth the probability of a given sequence with 1 *W*, and there are two-and-a-half times as many such sequences—$\binom{6}{2} = 15$ versus $\binom{6}{1} = 6$.

Continuing, let's list the probabilities:

$$P(3 \text{ winners in } 6) = \binom{6}{3}\left(\frac{1}{6}\right)^3\left(\frac{5}{6}\right)^3 \sim 0.054,$$

$$P(4 \text{ winners in } 6) = \binom{6}{4}\left(\frac{1}{6}\right)^4\left(\frac{5}{6}\right)^2 \sim 0.010,$$

$$P(5 \text{ winners in } 6) = \binom{6}{5}\left(\frac{1}{6}\right)^5\left(\frac{5}{6}\right) \sim 0.0006,$$

$$P(6 \text{ winners in } 6) = \left(\frac{1}{6}\right)^6 \sim 0.00002.$$

So: what is the probability of getting at least one winning ticket in 6? This clearly calls for the subtraction principle: the probability of getting at least one ticket is 1 minus the probability of getting *no* tickets—that is, the likelihood of the outcome *LLLLLL*, which as we said above was

$$P(LLLLLL) = \left(\frac{5}{6}\right)^6 \sim 0.335,$$

or just about 1 in 3. So the probability of getting at least one winner is about

$$P(\text{at least 1 winner in } 6) \sim 1 - 0.335 = 0.665,$$

or about 2 in 3.

The sharp-eyed reader may note that we didn't really need the fancy formula $P(W^aL^b) = (1/6)^a(5/6)^b$ to answer the question in this case. Don't worry, our effort will not have been wasted when we deal with the remaining two cases.

Moving on the case of 12 tickets, we can calculate the probabilities of each possible result similarly. For example, the likelihood of no winners is

$$P(0 \text{ winners in } 12) = P(L^{12}) = \left(\frac{5}{6}\right)^{12} \sim 0.112.$$

Likewise, there are $\binom{12}{1} = 12$ outcomes involving exactly one winner, so

$$P(1 \text{ winner in } 12) = \binom{12}{1}\left(\frac{1}{6}\right)\left(\frac{5}{6}\right)^{11} \sim 0.269,$$

and the probability of exactly two winners is

$$P(2 \text{ winners in } 12) = \binom{12}{2}\left(\frac{1}{6}\right)^{2}\left(\frac{5}{6}\right)^{10} \sim 0.296.$$

Just one more: the likelihood of exactly three winners is

$$P(3 \text{ winners in } 12) = \binom{12}{3}\left(\frac{1}{6}\right)^{3}\left(\frac{5}{6}\right)^{9} \sim 0.197,$$

and it's not hard to see that it continues to drop from this point.

We actually didn't need the last two calculations to answer Pepys' question: the probability of getting at least two winners is 1 minus the probability of getting either no or one winner, which is to say

$$P(\text{at least 2 winners in } 12) \sim 1 - 0.112 - 0.269 = 0.619,$$

a little lower than the probability of getting at least one winner in six.

As for the case of 18 tickets, we'll just do the necessary cases here: the probability of getting no winners is

$$P(0 \text{ winners in } 18) = P(L^{18}) = \left(\frac{5}{6}\right)^{18} \sim 0.038;$$

the probability of exactly one winner is

$$P(1 \text{ winner in } 18) = \binom{18}{1}\left(\frac{1}{6}\right)\left(\frac{5}{6}\right)^{17} \sim 0.135,$$

and the probability of exactly two winners is

$$P(2 \text{ winners in } 12) = \binom{18}{2}\left(\frac{1}{6}\right)^{2}\left(\frac{5}{6}\right)^{16} \sim 0.230.$$

It follows that the probability of getting three or more winners is just 1 minus the sum of these, that is,

$$P(\text{at least 3 winners in } 18) \sim 1 - 0.038 - 0.135 - 0.230 = 0.597.$$

Thus, the answer to Pepys' question is that at least 1 in 6 is the most likely.

Exercise 10.1.1. Pepys has a biased coin with a $1/6$ probability of coming up heads and a $5/6$ probability of coming up tails.

1. Which is the most likely outcome of flipping Pepys' coin: one heads in six flips, two heads in twelve flips, or three heads in eighteen flips?
2. Where have you seen this question before?

Exercise 10.1.2. Give a conceptual justification to why 1 win in 6 is more likely than 2 wins in 12 or 3 wins in 18 in Pepys' lottery.

Exercise 10.1.3. Suppose you have a fair coin, with equal probabilities of coming up heads or tails, but you want to use it to simulate a biased coin, with a $1/4$ probability of coming up heads and a $3/4$ probability of coming up tails. How might this be achieved?

10.2 BERNOULLI TRIALS

At this point, we need some terminology and a general formula to describe what we're doing. In general, suppose we have an event with two possible outcomes, called A and B. This could be a coin flip; or, it could be an event with more than two possible outcomes, where we lump together the possible outcomes into two classes A and B: for example, the event could be the roll of a die, with event A being a 6 and event B being any other number. Let's say the probability of A occurring is p, and the probability of B is $q = 1 - p$; if we were flipping a fair coin, with A heads and B tails, of course, we'd have $p = q = 1/2$; in the example of the single die roll we'd have $p = 1/6$ and $q = 5/6$.

We now repeat this event a number n of times, so that the outcome of the sequence of n repetitions can be thought of a word of n letters, all of which are either A or B. Moreover, we assume that these iterations are *independent*, meaning that the probabilities on each event in the sequence are the same—$P(A) = p$ and $P(B) = q$—regardless of the result of previous iterations.

Such a sequence of independent iterations of an event with two outcomes is called a series of *Bernoulli trials*. There are many questions we can ask about Bernoulli trials: what is the likelihood of outcome A occurring in exactly k of the n iterations; what we might expect the longest streak of As or Bs to be; the gambler's ruin problem described in the next section; and many others. For now, though, we'll focus on the first, which is the simplest.

In fact, the way we've already seen how to answer this in the case of Pepys' question applies in general. If, for example, n is 7 and we ask what is the likelihood of the specific outcome $ABAABBB$, the answer is

$$P(ABAABBB) = p \cdot q \cdot p \cdot p \cdot q \cdot q \cdot q = p^3 q^4,$$

and in general the likelihood of any specific outcome involving k As—that is, any sequence of As and Bs consisting at k As and $\ell = n - k$ Bs—is

$$P(A^k B^l) = p^k q^l.$$

Now, of all the 2^n possible outcomes, we know that the number that involve exactly k As and ℓ Bs is $\binom{n}{k}$ (equivalently, $\binom{n}{\ell}$), and *they all have the same probability of occurring.*

The likelihood of getting a total of exactly k As in n tries is thus the product $\binom{n}{k}p^k q^\ell$, a formula that we'll use many times and so deserves a box:

In n trials of an event where the outcome A occurs with probability p, the probability of having A occur exactly k times is:

$$P(\text{exactly } k \text{ As}) = \binom{n}{k}p^k(1 - p)^{n-k}.$$

Problem 10.2.1. Say you're playing a series of 10 hands of 5-card poker. What is the likelihood of being dealt the ace of spades exactly three times?

Solution. To answer this, we first need to know the likelihood p of getting the ace of spades on a given hand. This is not hard: you're dealt five cards at random out of the deck of 52; the probability that the ace of spades is among them is

$$p = \frac{5}{52}.$$

Now, given this, what are the probability of this happening exactly three times in a series of 10 hands? We could just plug into the boxed formula above, with $n = 10$ and $k = 3$, but we'll go through the logic one more time. Let A denote the outcome that you get the ace of spades on a given hand, and B the outcome that you don't. There are $\binom{10}{3} = 120$ sequences of As and Bs consisting of exactly 3 As and 7 Bs, and the probability that a given one of these will actually occur, on a sequence of 10 deals, is

$$P(A^3 B^7) = \left(\frac{5}{52}\right)^3 \left(\frac{47}{52}\right)^7.$$

Thus the probability that one of these sequences occurs is

$$120 \cdot \left(\frac{5}{52}\right)^3 \left(\frac{47}{52}\right)^7 \sim 0.0526. \qquad \square$$

Problem 10.2.2. Say you're taking a multiple-choice exam, consisting of 15 questions, with 5 possible answers for each one. The grading system is that you get 4 points for each correct answer, but 1 point is subtracted for each wrong one. If you answer all the questions randomly, what is the probability of winding up with a negative score? (Note: 0 is not a negative number.)

Solution. The first thing to figure out here is how many questions you have to get wrong to get a negative total. This is not difficult: if you get 4 times as many wrong answers as right ones—that is, 12 wrong and 3 right out of the 15 questions—you get a zero; to get a negative score, accordingly, you'd have to get fewer than 3 right.

That said, we have to calculate the probability of getting either 0, 1, or 2 questions right. Now we will just plug into the formula: the total number of trials is $n = 15$, and the probability of success on each is $1/5$, so the chances of getting 0, 1, or 2 right are, respectively,

$$P(0 \text{ right}) = \left(\frac{4}{5}\right)^{15} \sim 0.035,$$

$$P(1 \text{ right}) = \binom{15}{1}\left(\frac{1}{5}\right)\left(\frac{4}{5}\right)^{14} \sim 0.132,$$

and

$$P(2 \text{ right}) = \binom{15}{2}\left(\frac{1}{5}\right)^{2}\left(\frac{4}{5}\right)^{13} \sim 0.231.$$

Adding these up, we see that the probability of getting a negative score is

$$0.035 + 0.132 + 0.231 = 0.398,$$

or about 2 in 5. $\square$

   This example is based, as they say, on true events: one of us had a roommate who managed to get a negative score on both midterms and the final in multivariable calculus.

   Bernoulli trials can also help us compute the most likely outcome of rolling fair or unfair dice. For instance, suppose you roll 10 dice. Which is more likely, that you get one 6 or that you get two 6s?

   Now, if you were to approach the problem naively, you might say that on average the number of 6s you'll get in 10 dice rolls is $10/6 \sim 1.667$. The closest whole number to 1.667 is 2, so you might guess that the most common outcome would be to roll two 6s. That's not far off the mark, as we'll see in a moment, but it's not right: if we use our formula on page 137 to calculate the chances of each, we find that

$$P(\text{one } 6) = \binom{10}{1}\left(\frac{1}{6}\right)\left(\frac{5}{6}\right)^{9}$$

and

$$P(\text{two } 6\text{s}) = \binom{10}{2}\left(\frac{1}{6}\right)^{2}\left(\frac{5}{6}\right)^{8}.$$

Now, we could just multiply these out and see which one's larger. But in fact, if we compare the two expressions above, we can tell which one's bigger without hauling out the calculator. Put it this way: comparing the expressions

$$\left(\frac{1}{6}\right)\left(\frac{5}{6}\right)^{9} \quad \text{and} \quad \left(\frac{1}{6}\right)^{2}\left(\frac{5}{6}\right)^{8},$$

which represent the probability of *a given sequence* involving one 6 or two 6s occurring, we see that the latter is one-fifth of the former: we've replaced a $5/6$ in the left-hand expression with a $1/6$ in the right. On the other hand, there are more sequences involving two 6s than sequences involving just one; the numbers are

$$\binom{10}{1} = 10 \quad \text{and} \quad \binom{10}{2} = 45.$$

Thus, to go from P(one 6) to P(two 6s), we multiply by $45/10 = 4.5$, but divide by 5; the result is that P(two 6s) is slightly smaller than P(one 6).

Using this example as a model, let's try analyzing the general situation: we do n Bernoulli trials of an event having outcomes A and B with probabilities p and $q = 1 - p$, and ask what is the most likely number of As—that is, for what number k is the probability

$$P(\text{exactly } k \text{ As}) = \binom{n}{k} p^k q^{n-k}$$

the largest?

To answer this, we do what we did in the example above: we compare the quantities $\binom{n}{k} p^k (1-p)^{n-k}$ as k runs from 0 up to n. To start, compare the first two cases $k = 0$ and $k = 1$—that is,

$$P(\text{no As}) = \binom{n}{0} q^n$$

and

$$P(\text{one A}) = \binom{n}{1} pq^{n-1}.$$

The binomial coefficient $\binom{n}{1} = n$ is n times bigger than the binomial coefficient $\binom{n}{0} = 1$. On the other hand, the expression $p \cdot q^{n-1}$ is p/q times q^n. So: if p/q is bigger than $1/n$, then P(one A) is bigger than P(no As), and vice versa.

Similarly, lets compare the expressions

$$P(\text{exactly } k \text{ As}) = \binom{n}{k} p^k q^{n-k}$$

and

$$P(\text{exactly } k + 1 \text{ As}) = \binom{n}{k+1} p^{k+1} q^{n-k-1}.$$

As before, the second half of the latter expression—that is, $p^{k+1} q^{n-k-1}$—is p/q times the corresponding part $p^k q^{n-k}$ of the former. On the other hand, if we compare the binomial coefficients

$$\binom{n}{k} = \frac{n!}{k!\,(n-k)!} \quad \text{and} \quad \binom{n}{k+1} = \frac{n!}{(k+1)!\,(n-k-1)!}$$

we see that the latter is obtained from the former by converting the $k!$ in the denominator to a $(k+1)!$—that is, dividing by $k + 1$—and likewise converting the $(n - k)!$ in the denominator to an $(n - k - 1)!$, or multiplying by $n - k$. We conclude that

$$\binom{n}{k+1} = \frac{n-k}{k+1} \binom{n}{k}.$$

Combining these factors, we see that if q/p is smaller than $n-k/k+1$, then the probability P(exactly $k + 1$ As) will be bigger than P(exactly k As), and vice versa.

That's way too many letters. To get a sense of what's going on, we can express it qualitatively: In n Bernoulli trials with probability p of winning, the likelihood of winning exactly k times gets larger as k increases, up to the point where the ratio $n-k/k+1$ drops below $1-p/p$; from that point on the probability of winning exactly k times gets smaller as k increases. The peak—the number k of wins most likely to occur—will be at one of the two whole numbers on either side of the expected value np of the n trials.

This problem is just a little tougher than the examples considered above, in that we have to start by figuring out the probability of the individual event.

Exercise 10.2.3.

1. What is the probability of being dealt an ace in five cards?
2. What is the likelihood, in a series of seven poker hands, of being dealt an ace five times?

Exercise 10.2.4.   Spades is a four-player game in which each player is dealt 13 cards from a standard 52-card deck (and spades are trump, with the ace high).

1. What is the probability of being dealt an ace of spades in 10 hands of spades?
2. Is this more or less likely than the probability of being dealt the ace of spades in 10 hands of 5-card poker?

Exercise 10.2.5.   Now Pepys' has made a loaded six-sided dice, for which the probabilities of rolling a 1 or rolling a 2 are each $1/4$, the probability of rolling a 3 is $1/6$, and the probabilities of rolling a 4, rolling a 5, or rolling a 6 are each $1/9$.

1. Check that these probabilities sum to 1.
2. After 10 dice rolls, what is the probability that exactly half of the rolls result in a 4, 5, or 6?
3. After 10 dice rolls, what is the probability that at least half of the rolls result in a 4, 5, or 6?

Exercise 10.2.6.   Let's say Nathan and Carl are playing dice: they roll one die; if it comes up 6, Nathan pays Carl $5, and if it comes up anything but 6, Carl pays Nathan $1. Let's say they play 11 times. What is the probability that Nathan will be ahead at the end?

10.3   GAMBLER'S RUIN

Here's the classic gambler's ruin setup: you're playing roulette at a casino; you have $1,000. You've decided that you're going to keep playing until either you have $2,000 or you're broke.

Your gambling shows a distinct lack of imagination: you always bet on black. That is, you decide how much to bet on each spin of the wheel: if it comes up black, you win that amount; if not, you lose that amount. The roulette wheel at this casino comes up black roughly 48% of the time.

As you sit down at the table, you ponder two possible strategies. One is simply to bet the entire $1,000 in one bet. The other is more conservative: bet $1 at a time. Which strategy gives you better odds of walking away with two grand? We know that if you bet it all on one spin, your chances of winning are 48%. What is the probability of winning, and what is the probability of going broke, if you bet $1 at a time?

This is the gambler's ruin problem, or, as it's called in the trade, a random walk with absorbing barriers. The more picaresque name may suggest that your odds are not good. But it's worth pausing a moment to take a guess: what is the actual likelihood of success? It'll be a while before we have the answer, so write your answer down on a piece of paper.

Our approach to introducing new ideas and techniques up to now has been to start with examples, and then go to the general case. That doesn't work here: even relatively simple instances of gambler's ruin—with a stake of a few dollars—can get complicated.

In fact, this is one of those instances, not uncommon in mathematics, where it's easier to solve a more general problem. So: we're going to take away your $1,000, and give you instead a dollars; you're going to play until either you have b dollars, or you're broke (that part doesn't change). The probability of coming up black isn't 48% any more; it's p, which we'll think of as less than $1/2$. But hey, you still bet $1 at a time. We ask: what is the probability—which we'll call $P(a)$ since, as we'll discover momentarily, the value depends strongly on the opening stake of a dollars—of winding up with the b dollars?

Let's start with the obvious: if you're broke—that is, $a = 0$—the game's over and you've already lost, so $P(0) = 0$. Likewise, if $a = b$—you start with b dollars—you've already won; so $P(b) = 1$. What about all the intermediate values of a, from 1 up to $b - 1$? Well, in those cases it's time to bet. Now, of course we don't know the outcome of our first bet—if the first spin is going to come up black or not. But *conditional probability allows us to deal with this ambiguity*. There are two possible outcomes of the first spin:

- The fraction p of the time, the wheel will come up black and we'll now have $a + 1$ dollars. Our probability of success in this case is $P(a + 1)$.
- The other $q = 1 - p$ of the time, we lose the initial bet; in these cases, our stake is reduced to $a - 1$ dollars and our probability of success is $P(a - 1)$.

This, if you think about it, is exactly the situation for conditional probability: we don't know how the first bet's going to go, but we know the probabilities in either case. Applying the basic conditional probability formula from page 112, we see that

$$P(a) = p \cdot P(a+1) + q \cdot P(a-1).$$

In other words, the probability $P(a)$ of success if we start with a stake of a dollars is a *weighted average* of $P(a + 1)$ and $P(a - 1)$.

To sum up, here's what we know about $P(a)$: we know that $P(0) = 0$, $P(b) = 1$, and $P(a) = p \cdot P(a+1) + (1-p) \cdot P(a-1)$ for all the values of a in between. All we have to do is to solve these $a + 1$ equations! Luckily, there's a trick that allows us to do exactly that. The trick is to ignore the common-sense conditions $P(0) = 0$ and $P(b) = 1$, and just find some solutions of the equations

$$P(a) = p \cdot P(a+1) + (1-p) \cdot P(a-1).$$

In fact, it's not hard to do that: for example, the function $P(a) = 1$ satisfies them all. Another one that works is

$$P(a) = \left(\frac{q}{p}\right)^a.$$

Check it out: if $P(a) = (q/p)^a$, then

$$p \cdot P(a+1) = p \cdot \left(\frac{q}{p}\right)^{a+1} = \frac{q^{a+1}}{p^a} = q \cdot \left(\frac{q}{p}\right)^a$$

and likewise

$$q \cdot P(a-1) = q \cdot \left(\frac{q}{p}\right)^{a-1} = \frac{q^a}{p^{a-1}} = q \cdot \left(\frac{q}{p}\right)^a,$$

and since $p + q = 1$, when we add these two together, we get

$$p \cdot P(a+1) + q \cdot P(a-1) = q \cdot \left(\frac{q}{p}\right)^a + q \cdot \left(\frac{q}{p}\right)^a$$
$$= \left(\frac{q}{p}\right)^a$$
$$= P(a),$$

as desired.

More generally, any multiple of either of the functions $P(a) = 1$ and $P(a) = (q/p)^a$ also works, as does any sum of these; so all we have to do now is find some sum of multiples of these two functions that satisfies the conditions $P(0) = 0$ and $P(b) = 1$. We won't go through the process of finding such a sum, but will just give you the answer: the combination that works is

$$P(a) = \frac{-1}{\left(\frac{q}{p}\right)^b - 1} - \frac{1}{\left(\frac{q}{p}\right)^b - 1} \cdot \left(\frac{q}{p}\right)^a,$$

as you can check for yourself. Cleaning up, we can rewrite this as

$$P(a) = \frac{\left(\frac{q}{p}\right)^a - 1}{\left(\frac{q}{p}\right)^b - 1}.$$

Finally, we can write this in a simpler form if we introduce one more letter: let $s = q/p$ be the ratio between the probability q of losing a given bet and the probability p of winning it. We can then write our answer as

$$P(a) = \frac{s^a - 1}{s^b - 1},$$

and this is our probability of succeeding in the gambler's ruin situation.

To apply this is simple and will illuminate how Las Vegas stays in business. In the classic gambler's ruin problem we started with, we start with a stake of \$1,000, so $a = 1,000$; and we play until we either go broke or have \$2,000, so $b = 2,000$. The chances of winning a given bet is 48%, and the likelihood of losing 52%, so the ratio $s = q/p$ is

$$r = \frac{52}{48} = \frac{13}{12},$$

and so by our general formula, the probability of success is

$$\frac{\left(\frac{13}{12}\right)^{1,000} - 1}{\left(\frac{13}{12}\right)^{2,000} - 1}.$$

Hauling out our calculator—and it better be a good one, with scientific notation, because a number like $(13/12)^{2,000}$ has a *lot* of digits—we see that the probability of winning is

$$\frac{\left(\frac{13}{12}\right)^{1,000} - 1}{\left(\frac{13}{12}\right)^{2,000} - 1} \sim \frac{\left(\frac{13}{12}\right)^{1,000}}{\left(\frac{13}{12}\right)^{2,000}}$$

$$= \left(\frac{13}{12}\right)^{-1,000}$$

$$\sim 0.00000000000000000000000000000000000173,$$

or 1.73×10^{-35} in exponential notation.

Put it this way: Suppose you went to Las Vegas at the dawn of our universe and performed this experiment—starting with \$1,000 each time and betting \$1 on black until you either lost your stake or won an additional \$1,000—once every day. The odds are, *you still wouldn't have succeeded in reaching* \$2,000 *even once.*

And that, in one simple formula, is how Las Vegas stays in business, and can even afford to give you all that food and drink for free as well as build tacky replicas of Venetian palaces. You go there to play roulette, having been told that that's one of the least house-favorable games on offer (it is). You avoid all the obvious pitfalls: you limit your stake, and you decide in advance that if you reach a certain sum in winnings you'll just walk away. You even decide to play conservatively and wager only \$1 each time. You think, after all that, that you have a 48% chance of winning. You don't. Playing as we've described, *you have, in effect, absolutely no chance of winning.*

You may be skeptical of this conclusion: after all, people do occasionally win at Las Vegas, right? This is certainly true; the key point here was the decision to bet only \$1 at a time. After all, if you bet your whole stake of \$1,000 on one spin of the wheel, and walked away if you won, you would indeed have a 48% chance of success. The point is, the smaller the bet each time, the longer you play; and the longer you play, the more certain it is that you'll lose. In other words, your mistake was thinking that betting only \$1 at a time was conservative: if you really want to maximize your chances of solvency, you'll bet it all at once!

To make this clearer, let's work out the probabilities using some intermediate strategies:

Problem 10.3.1. Say as before you go to Vegas with \$1,000, play roulette, and bet on black until you either have \$2,000 or nothing; say that the probability is again 48% of winning on any given spin. What is your likelihood of success if:

1. you bet \$10 at a time;
2. you bet \$100 at a time; or
3. you bet \$250 at a time.

Solution. To apply our general formula, just think in terms of "chips" rather than "dollars" specifically. For example, if you want to bet \$10 at a time, you convert your stake of \$1,000 into 100 ten-dollar chips, which you'll bet one at a time until you either have none left or have 200. Thus in this situation we have $a = 100$ and $A = 200$, and the chances of success are

$$\frac{\left(\frac{13}{12}\right)^{100} - 1}{\left(\frac{13}{12}\right)^{200} - 1} \sim \frac{\left(\frac{13}{12}\right)^{100}}{\left(\frac{13}{12}\right)^{200}}$$

$$= \left(\frac{13}{12}\right)^{-100}$$

$$\sim 0.000334,$$

or about one in 3,000 tries. Much better!

Likewise, if we want to bet \$100 at a time, we get a stack of 10 hundred-dollar chips; now $a = 10$ and $A = 20$, and our probability is

$$\frac{\left(\frac{13}{12}\right)^{10} - 1}{\left(\frac{13}{12}\right)^{20} - 1} \sim 0.2474,$$

or almost 1 in 4. Finally, if we convert our stake to four \$250 chips, we have $a = 4$, $A = 8$, and our probability is

$$\frac{\left(\frac{13}{12}\right)^{4} - 1}{\left(\frac{13}{12}\right)^{8} - 1} \sim 0.4206.$$

Finally, we're in range of the posted 48% probability on each spin—but we'd still be better off betting the whole bundle at once. $\square$

Here's another example of this formula, this time with baseball.

Problem 10.3.2. Frustrated by the inability of the World Series, as currently constituted, with the first team to win four games declared the winner, to go beyond seven games, Major League Baseball announces a change in format: the Boston Red Sox and the Chicago Cubs will play until one team is up by four, that is, has won four more games than the other. Assume that the Red Sox (being the better team) will win games against the Cubs on average 60% of the time. What is the probability that the Cubs will win the Series?

Solution. The first thing is to see that this is a version of the gambler's ruin problem. Put it this way: suppose you bet on the Cubs (not that anyone ever bets on professional sports or anything). You start with four dollars, and bet one dollar at even money on each game—if the Cubs win, you get a dollar; if the Red Sox win, you lose a dollar. If you go broke, that means the Red Sox have won four more games than the Cubs, and so won the Series. On the other hand, if your stake ever gets to \$8, that means the Cubs have won four more times than they've lost, and so won the Series. So this is really just the gambler's ruin problem all over again, this time with $a = 4$, $A = 8$, and probabilities $p = 2/5$ and $q = 1 - p = 3/5$. The key ratio is then

$$s = \frac{q}{p} = \frac{3/5}{2/5} = \frac{3}{2},$$

and our formula says that the likelihood of the Cubs winning is

$$\frac{\left(\frac{3}{2}\right)^{4} - 1}{\left(\frac{3}{2}\right)^{8} - 1} \sim 0.165,$$

or about 1 in 6. $\square$

Note that in the new format, even though the Red Sox are by hypothesis the better team, there's a 1-in-6 chance the Cubs will win the Series. The following question asks you to compare this probability with the traditional seven-game series format.

Exercise 10.3.3.

1. What is the probability of the Cubs winning in the traditional format, where the teams play until one team wins four games?
2. Which format—the traditional first-team-to-win-four-games format, or the new one described above—is better at sorting out who's the better team?

Exercise 10.3.4. In tennis, if two players are tied 3-3 after six points they continue to play until one player is ahead of the other by two points; that player wins the game. Suppose that Serena and Venus are playing tennis, and that Serena will win a point over Venus on average 2/3 of the time. If they're tied at 3-3, what is the probability that Venus will win?

Exercise 10.3.5. Main Street in Middletown runs north-south a total of ten blocks, with a restaurant at each end—Anna's Taqueria on the north end, and Hokkaido Ramen Santouka at the south end. You're standing in the middle, five blocks from each, and in need of a meal. You can't decide which way to go, so you decide to leave it up to the fates. Specifically, you decide you'll roll one die; if it comes up between 1 and 4 you'll walk one block north, and if it comes up 5 or 6 you'll walk one block south. You repeat this until you arrive at one or the other restaurant. What is the probability you'll wind up at Hokkaido Ramen Santouka?

11 Geometric probability

All of our discussions of probability thus far have been limited in one significant aspect: in each case, there were only finitely many possible outcomes of a given event. But that's very restrictive, in some ways. In most of life's processes, the inputs and outputs are things that can vary continuously—time, location, distance, height, and weight for example. In this unit, we'll see how to treat problems involving continuously varying quantities.

11.1 COIN TOSSING AT THE CARNIVAL

There is a classic coin tossing carnival game. A table is marked off by a grid of lines one inch apart. You toss a quarter, which has a diameter of $3/4$ of an inch, onto the table. If it lands entirely in a square, you win; if it crosses one of the lines of the grid, you lose. What are your chances?

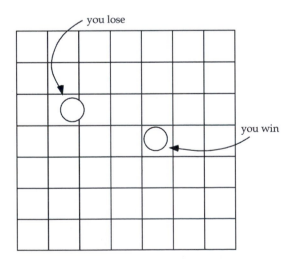

The first thing to say here is that this is an example of a problem where the possible outcomes vary in a continuum. In this case, the position of the coin after landing—that is, the location of its center—could be any point on the table.

Secondly, we assume that you have no hope of actually aiming the thing—in other words, the probability of the center of the coin landing in any given region is proportional to the area of that region. (We'll have more to say about that assumption in the discussion of the next example, in Section 11.2.) Now, the rules are that you lose the game when your coin touches any of the lines of the grid, which is to say when the center of coin lies within its radius (here, that would be 3/8 of an inch) of any of the lines. What we have to do, accordingly, is to draw the locus of points on the table that are within a distance of 3/8 of the lines, and find its area as a fraction of the total area of the table.

The next thing to observe is that we can do this square by square: within each square, the locus of points within a distance of 3/8 of the sides of the square represents a certain fraction of the area of that square; and since the same pattern is replicated in each square, that fraction is the fraction of all points on the tabletop with that distance of a line. So let's look at a single square here, and see what proportion of its area lies within a distance of 3/8 of its sides:

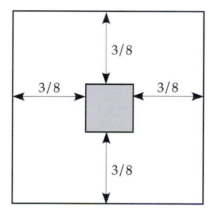

Here the shaded square represents the favorable outcomes—the points lying a distance of 3/8 or more away from the sides, so that if the center of the coin lands in that region, you win. Unfortunately for you, this is a square whose sides have length

$$1 - \frac{3}{8} - \frac{3}{8} = \frac{1}{4}$$

and which therefore has area 1/16 of a square inch. In other words, you win one-sixteenth of the time; you lose 15 times out of 16.

Our next game is manifestly not a classic carnival game, unless carnivals have started traveling with random-number generators. The deal on this one is that you get two random numbers between 0 and 1; as in all the other examples here, we assume that the two are independent, and that each one is as likely to come from any given portion of the line between 0 and 1 as any other portion of the same length. You win if the sum of the two numbers is bigger than 1.5; otherwise, you lose. What are your chances?

Well, once more this is just a matter of drawing a square representing all possible outcomes—that is, all pairs of number x and y between 0 and 1—and finding the area of the region consisting of points (x, y) such that $x + y \geq 1.5$. This is straightforward; the line $x + y = 1.5$ cuts off just one corner of the square of possible outcomes, as shown:

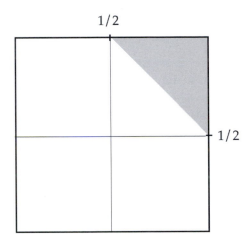

1/2

1/2

The area of this triangle being 1/8 (if you don't happen to remember the formula for the area of a triangle, this one is one half of a $1/2 \times 1/2$ square), the probability of you winning is just one in 8.

One last word before we leave the carnival and look at some other instances of geometric probability. Of course carnivals don't have random number generators. But they do have things like giant wheels, with the numbers between 1 and 100 evenly spaced around the rim; they might bet you that if you spin the wheel twice, the sum of the numbers you get will be more or less than 150. This is at first glance not a problem that belongs in this section: it involves only a finite number ($100^2 = 10{,}000$) of possible outcomes, all equally likely. But think about how you'd go about counting the number of winning outcomes—it could get exhausting, and after all you're at the carnival to have fun. It's actually easier, in a case like this, to approximate the probabilities in the finite problem by calculating the continuous version!

Exercise 11.1.1. You play a simple game where you're given two numbers between 0 and 1 at random, as in the last example. Your payoff is \$1 when $1 < a + b \leq 1.5$ and \$3 when $1.5 < a + b$. What is the expected value of playing this game?

11.2   AT THE DINING HALL

Let's start with an example:

Problem 11.2.1. Let's say the cafeteria is open for dinner between the hours of 5:30 and 7:30. Your schedule being fragmented, you tend to arrive at the cafeteria at a random time in that interval, and you stay for half an hour eating. Say the same is true of a friend of yours. What is the probability that you'll see each other on a given day?

As in the coin-toss problem, the unknown quantities—the times you and your friend arrive at the cafeteria for dinner—vary in a continuum; they can be any moments between 5:30 and 7:30. If we wanted to use the techniques of finite probability, we'd have to break down the set of possible times between 5:30 and 7:30 into a finite number of categories, or ranges: you could, for example, break the two hours down into 30-minute increments, and say that you are equally likely to arrive in any of the four ranges 5:30–6, 6–6:30, 6:30–7, and 7–7:30. The problem here is that knowing which of the four ranges you're in, and likewise for your friend, isn't always enough information to determine whether the two of you see each other: if, for example, you know that you arrived between 6 and 6:30, and your friend between 6:30 and 7, you might overlap or you might not.

Happily, though, there is a way of dealing with this problem. It starts with saying a little more precisely what we mean by "you arrive at the cafeteria at a random time between 5:30 and 7:30." We don't mean to be tiresome about it, but it's an important point: we can't fudge this by saying "you're no more or less likely to arrive at one time or any other." In fact, we can't ascribe any probability at all to the event that you arrive at a given moment, since there are infinitely many instants between 5:30 and 7:30.

So what do we mean by a random time? About the only thing we can mean is that *the probability that you'll arrive in a given interval is equal to the probability that you'll arrive in any other interval of the same duration.* So, for example, the likelihood of your arriving between 5:50 and 5:55, say, are the same as the likelihood of arriving between 6:35 and 6:40, or between 7:02 and 7:07, or any other five-minute period.

Now, if we assume that, it follows that we much have an equal chance of arriving between 5:30 and 6:30 or between 6:30 and 7:30; since we have to arrive in exactly one of these ranges, the probability of each must be exactly 1/2. It follows also that the likelihood of our arriving in any given hour-long interval—say, between 5:37 and 6:37—must be 1/2 as well. By the same token, the probability of our arriving in any half-hour interval must be 1/4; and, in general, the probability of our arrival in any interval is one-half the length, in hours, of that interval.

Let's give ourselves a break here and represent the times graphically:

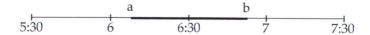

If we think of the entire time line between 5:30 and 7:30 as having length 1—so that each time in between corresponds to a number between 0 and 1—then the probability that your time t of arrival is between two given times represented by the numbers a and b is just $b - a$: in symbols,

$$P(a \leq t \leq b) = b - a.$$

Now, if we represent your time of arrival as a point on the line segment $[0, 1]$—that is, as a number t between 0 and 1—then we can represent your friend's time of arrival as a number s between 0 and 1. We can then view the pair of times t and s as a point in a square of unit side length:

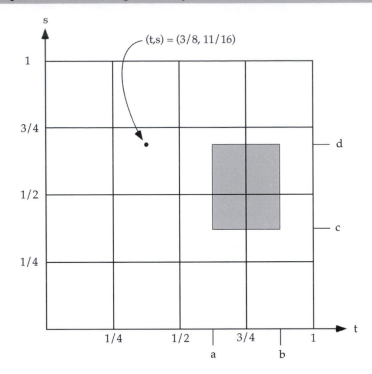

Next, we make a crucial assumption: that your time of arrival and your friend's are independent events. (In fact, we can use the calculation below to test the hypothesis of independence, as we'll see in the "How to tell if you're being stalked" section below.) In that case, the probability that your arrival time is between a and b and your friend's arrival time is between c and d is just the product of the probabilities of the two individual events: that is,

$$P(a \leq t \leq b \text{ and } c \leq s \leq d) = P(a \leq t \leq b) \cdot P(c \leq s \leq d)$$
$$= (b-a)(d-c).$$

Now, the product $(b-a)(d-c)$ is just the area of a rectangle with side lengths $b-a$ and $d-c$. In other words, the probability that the point (t, s) lies in the shaded rectangle in the picture is equal to the area of that rectangle. More generally it's the case that, if your time of arrival and your friends are independent, then *the probability that (t, s) falls in any region in the square is equal to the area of that region.*

And that's what we need to solve the problem. To say that you and your friend's dinners overlap is just to say that your arrival times differ by a half hour or less. Since a half hour is one-quarter of the time between 5:30 and 7:30, this is represented by the equation

$$|t - s| \leq \frac{1}{4},$$

or, geometrically, that the point (t, s) is within a distance of $1/4$ of the diagonal in the vertical (or, equivalently, horizontal) direction. The points (t, s) that satisfy this condition are represented by the shaded area in this figure:

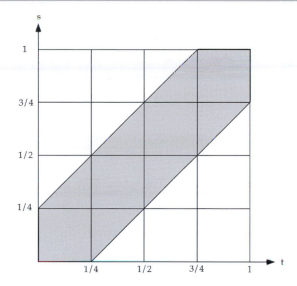

So the question is, what's the area of this shaded region? That's easy enough: the grid we've drawn breaks the region up into 4 squares, each $1/4$ by $1/4$, plus 6 isosceles right triangles with side length $1/4$. The squares have area $1/16$, and the triangles half that, or $1/32$; the total area is thus

$$4 \cdot \frac{1}{16} + 6 \cdot \frac{1}{32} = \frac{14}{32} = \frac{7}{16} = 0.4375,$$

and this is the probability $P(\text{overlap})$ that you and your friend will see each other at dinner.

We can combine this sort of analysis with the notion of conditional probability. For example, consider the question: what is the probability that you and your friend will overlap, assuming that you arrive between 5:30 and 6? In symbols, what is $P(\text{overlap assuming } 0 \le t \le 1/4)$?

To answer this, we basically restrict our attention to the rectangle R of points (t, s) with $0 \le t \le 1/4$:

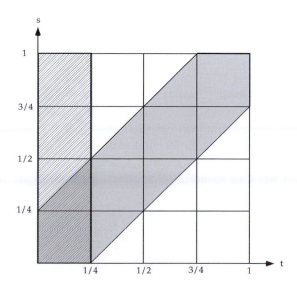

If we assume your arrival time t is in the interval between 0 and 1/4, that means the pair (t, s) of arrival times is a random point in the rectangle R. The probability that you and you friend will see each other is then

$$P(\text{overlap assuming } 0 \leq t \leq \frac{1}{4}) = \frac{\text{area of shaded part of rectangle } R}{\text{area of } R}.$$

Now, the area of the shaded portion of R is 3/32 while the area of R is 1/4 = 8/32; so we have

$$P(\text{overlap assuming } 0 \leq t \leq \frac{1}{4}) = \frac{3}{8}.$$

Now here are some questions for you to answer:

Exercise 11.2.2. Continuing in the previous setup:

1. What is the probability that you arrive before 6, *assuming that you and your friend see each other?*
2. What is the probability that at least one of you and your friend will arrive before 6?
3. What is the probability that at least one of you and your friend arrives before 6, assuming that you see each other?
4. Say the cafeteria serves lobster, but they run out around 6. You and your friend agree that if either of you arrives before 6, the first one to arrive will get two so you'll each have one—assuming that the second one arrives while the first is still there. What is the probability that at least one of you will arrive before 6 *and* that you'll see each other?

Exercise 11.2.3. Returning to the original problem, once more let's say a cafeteria is open for dinner between 5:30 and 7:30 and you and a friend independently arrive at random times during that interval. You stay for 30 minutes upon arrival but now suppose your friend is a fast eater and finishes dinner in 15 minutes. Now what are the chances your dinners will overlap?

11.3   HOW TO TELL IF YOU'RE BEING STALKED

Let's say that, as in the last series of problems, you arrive at the cafeteria at random times between 5:30 and 7:30. You notice at one point that you're seeing a lot of this one other person—it seems like almost every time you eat dinner, you see them. You decide to keep track, and over the course of the next 10 days you see them 8 times at dinner. Can this be a random occurrence, or are you being stalked?

We can start by calculating the likelihood of seeing this person at dinner 8 or more times in 10 days, assuming that their arrival times are in fact random. We have all the tools to answer this: assuming that their arrival times are random, the chances of running into them on a given day are 7/16. We can treat the 10 days as a series of Bernoulli trials, in which case as we've seen the chances of seeing them exactly 8 times out of 10 would be

$$\binom{10}{8}\left(\frac{7}{16}\right)^8\left(\frac{9}{16}\right)^2 \sim 0.019.$$

Likewise, the chances of seeing them exactly 9 or 10 times are, respectively,

$$\binom{10}{9}\left(\frac{7}{16}\right)^9\left(\frac{9}{16}\right) \sim 0.0033$$

and

$$\left(\frac{7}{16}\right)^{10} \sim 0.00025.$$

Adding these up, we see that *assuming that the other person's arrival times are random, there is only a 1-in-50 chance that you would see them at dinner 8 or more times in 10 days.*

So, does this mean there's only a 1-in-50 chance that their arrival times are random? No, no, no! Wait—that's not emphatic enough. Let's try again: NO, NO, NO!!! This is exactly the issue we raised in conjunction with Bayes' theorem. What we have established so far is that the probability of at least 8 sightings in 10 days, *assuming random arrival times*, is about 0.02. To put it more succinctly, let A be the event that you see this person 8 or more times in 10 days, and let B be the event that their arrival times at the cafeteria are random. What we know then is that

$$P(A \text{ assuming } B) \sim 0.02,$$

or about 2%. So: it's unlikely that you'd see this person that often, if their arrival times were indeed random. What we're actually asking, though, is something different: we're asking, what is the probability that their arrival times are random, given that you've seen then 8 or more times in 10 days? In other words,

$$\text{what is } P(B \text{ assuming } A)?$$

And these, as we've said repeatedly, are not the same thing. According to Bayes' theorem, in fact, in order to relate the two we'd need to know things like $P(B)$, or, equivalently, $P(\text{not } B)$—that is, what were the odds that you were being stalked independently of this observation. In other words, are stalkers a commonplace part of your life?

In sum: the fact that you see this person as often as you do may be significant—as a rule of thumb, statisticians view occurrences with less than a 5% probability as significant—but it certainly doesn't mean that the probability you're being stalked is 98%.

11.4 QUEUING THEORY

Here's a similar sort of problem. Suppose you're the manager of a remote post office branch which employs just one clerk. You probably don't need more—you don't get a lot of customers at this post office—but you're concerned that with only one clerk, some of your customers may have to wait for service, especially during relatively busy times like lunch hour. You observe the traffic over a period of time, and you observe that the post office has four customers who come in at midday: one who comes in every day at noon: one who comes in every day at 1, and two others who come in at independent random times between 12 and 1. The transactions with each of these customers take 10 minutes. Since no one at all shows up between 11 and 12, the teller is always available right at 12:00.

The question is, what is the probability that one or more of the customers will have to wait?

We start, we're afraid, by naming the four customers. The one who always arrives at 12:00 we'll call Early Bird (since his time of arrival is fixed and known in advance, his name won't come up that much). Likewise, the one who always shows up at 1:00 will be Late Bird. And the two regular customers who arrive at random times during the hour we'll call A and B.[1] We'll let a be the time in hours after 12 that A shows up, and similarly b the time that B shows up.

Now we draw a square representing possible arrival times (a, b) for A and B. Here's a picture:

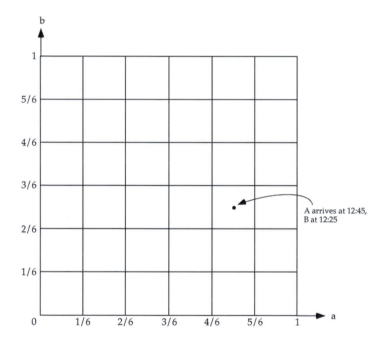

Here, the sides of the square represent the hour between 12 and 1, marked off into six increments of $1/6$ of an hour, or 10 minutes. As before, we make the sides of the square of length 1, so that as before the probability that the pair (a, b) of arrival times lies in a given region in the square is equal to the area of that region.

The next step is to mark off the possible arrival times (a, b) corresponding to scenarios where one or more of the customers will have to wait. To start, Early Bird never has to wait, since the teller is always available when he shows up at 12:00. If A arrives before 12:10, she'll have to wait; and likewise if she arrives after 12:50, then Late Bird will have to wait; so we mark off the regions in the square corresponding to these possibilities:

---

[1] This is *not* to say that A is necessarily the first person to show up and B the second; they are real human beings and they would like you to know that, despite their lackluster names, they are defined by more than when they happen to show up at the post office. As we said, their arrival times are independent; half the time B will show up before A and half the time vice versa.

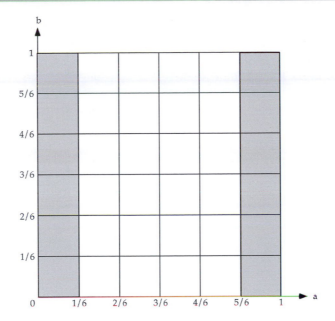

The same is true for B as well—if she arrives before 12:10 or after 12:50, someone's going to have to wait—and we mark off those outcomes as well:

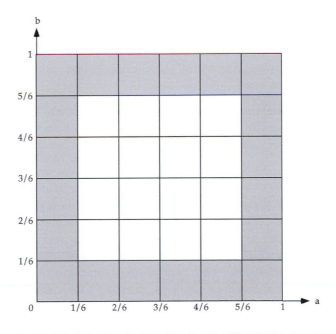

Finally, if A and B arrive within 10 minutes of each other, one of them will have to wait, so we mark off points (a, b) with a and b within 1/6 of each other, that is, within 1/6 of the diagonal either horizontally or vertically. Here's what we're left with:

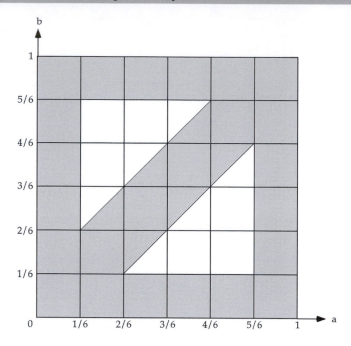

Finally, we're ready to answer the question of how likely it is that someone will have to wait: this is just the area of the shaded portion of our square. To calculate it, it might be easier just to find the area of the rest—the locus of (a, b) corresponding to outcomes where no one has to wait. This region consists of two triangles, that fit together to form a $1/2 \times 1/2$ square; so the probability that no one will have to wait is $1/2 \cdot 1/2 = 1/4$, and the likelihood that someone will have to wait is $1 - 1/4 = 3/4$, or three in four.

Exercise 11.4.1. If transactions took only 5 minutes instead of 10, what would be the probability that someone would have to wait?

Exercise 11.4.2. Suppose now that transactions take only 5 minutes, but B is old friends with the post office clerk and, what with all the chatting, spends 15 minutes at the window. As always, what is the probability that someone will have to wait?

Before we move on, we should mention that this could be viewed as an extremely simple example of a more general problem, belonging to an area called *queuing theory*. It is, naturally enough, concerned with what happens when a number of people show up at random times, but where the number of people may be large, and the total number of people is unknown. It's relevant not only to service businesses, as in the example we just did, but also to things like phone networks: if people make calls at random times, how large does the bandwidth of the network have to be to have, say, only a 5% chance of an overload occurring during a given call?

PART III

Probability at large

12 Games and their payoffs

The reader who's arrived at this point in the book might reasonably complain that we've learned to solve problems only in an artificially limited environment: in situations where there are a small number of undetermined outcomes, and a small number of possibilities for each. A flip of a coin, for example, has only two possible outcomes, and the sort of problems we've dealt with up to now have involved only a relatively small number of coin flips; so we can simply list the possible results and calculate the probability of each without too much wear and tear on our calculators.

But out in the world, life is rarely that simple. Consider, for example, an election with two candidates (Tracy and Paul, of course) running for president of the United States. We can think of each individual voter as a (possibly biased) coin, whose outcome is either a vote for Tracy or a vote for Paul. But now we're not talking about 5, or 10, or 20 "coin flips;" we're talking about 100,000,000. Even with the massive computing power we now possess, writing out the possible results of 100,000,000 coin flips and calculating the probability of each is a daunting prospect.

Fortunately, there is a way of dealing with this. In many situations where a problem in probability is so complex that we can't solve it simply by enumerating outcomes, there is a remarkable tool that allows us to at least approximate the answer. This tool, which we'll introduce in the next chapter, is called the *normal distribution*. We'll see that when we carry out a large number of Bernoulli trials, the results tend to approximate the normal distribution. This distribution, which is both an important computational tool and a key concept in the subject we call "probability at large," is central to many questions in probability and statistics.

In this chapter we'll introduce a mathematical model which will allow us to reason about the two situations considered just now. To motivate and illustrate these new ideas, let's start with a question:

Problem 12.0.1. Suppose you have a fair coin.

1. If you flip it 10 times, what are the odds you'll get 6 or more heads?
2. If you flip it 100 times, what are the odds you'll get 60 or more heads?
3. If you flip it 1,000 times, what are the odds you'll get 600 or more heads?

We'll answer all three questions in the course of this unit, and indeed the answer (given on page 176) will be striking (as always, you should take a moment out now and take a ballpark guess for each of the three questions—we promise it'll be instructive). But for now the main thing we want you to think about is: how would you go about finding the answer?

In the first question, we've already seen how to do this: the odds of getting exactly k heads out of 10 flips is

$$P(\text{exactly } k \text{ heads in 10 flips}) = \frac{\binom{10}{k}}{2^{10}},$$

so the answer to our first question would be the sum

$$P(\text{at least 6 heads in 10 flips}) = \frac{\binom{10}{6}}{2^{10}} + \frac{\binom{10}{7}}{2^{10}} + \frac{\binom{10}{8}}{2^{10}} + \frac{\binom{10}{9}}{2^{10}} + \frac{\binom{10}{10}}{2^{10}}$$

$$= \frac{\binom{10}{6} + \binom{10}{7} + \binom{10}{8} + \binom{10}{9} + \binom{10}{10}}{2^{10}}$$

$$= \frac{386}{1,024} \sim 0.377.$$

In principle, we could do the same thing to answer the second and third questions as well—but for the second question, we'd have to add up the 41 binomial coefficients of the form $\binom{100}{k}$ for $k = 60, 61, \ldots, 100$; for the third we'd have to add up 401 binomial coefficients of the form $\binom{1000}{k}$. Before we launch into such a massively tedious undertaking, we should ask: is there a better way to get the answer, at least approximately?

12.1   GAMES AND VARIANCE

To start, we want to set up a theoretical framework that will include virtually all the examples we've dealt with so far. There are lots of words we could use, ranging from the prosaic to the fairly fancy ("probability distribution"); we'll stick, as usual, to the former, and call the basic situation we'll be dealing with a *game*.

A game will consist of three things:

- An *event*, with a finite number of possible outcomes; if forced to, we'll call the possible outcomes $P_1, P_2, \ldots, P_k$, though usually there'll be more descriptive terms. (The case of a continuum of possible outcomes, which we discussed in Chapter 11, can be included as well—but only if we involve integral calculus, which we're not going to do.) The outcomes are what we say they are: if the event, for example, is the roll of a single fair die, we can declare the possible outcomes to be the numbers between 1 and 6; or, if we're interested solely in whether or not the die comes up 6, we can declare two outcomes, "6" and "not 6."
- A *probability* p_i associated to each of the possible outcomes P_i, meaning the fraction of the time the outcome P_i will occur. These numbers $p_1, \ldots, p_k$ will be numbers between 0 and 1, and the one thing we know in general is *they must add up to 1*. Finally:
- Associated to each possible outcome P_i is a numerical value, or *payoff*, which can be positive or negative. We've seen lots of probability problems that don't have explicit payoffs; in these cases—where we're just concerned with how frequently a given one of the outcomes occurs—we typically assign the outcome we're interested in a payoff of 1 and the others 0.

We'll often denote a game by a symbol, like G.

Examples of this situation include the flip of a coin, fair or unfair; a roll of one die or several dice; the deal of a hand of cards from a standard deck; the selection of a random voter in a poll—basically, virtually everything we've looked at so far (again with the exclusion of Chapter 11) can be put in this framework. In some of these instances, we specified payoffs associated to the various possible outcomes; in others, where we asked only for the likelihood of a given outcome, you could imagine associating a payoff of 1 to that outcome and 0 to all the others.

We have already introduced one key quantity associated to a game G: the *expected value* ev(G), the average of all the payoffs weighted according to their probability: in mathematical terms,

$$\text{ev}(G) = p_1 a_1 + p_2 a_2 + \cdots + p_k a_k.$$

Briefly, the expected value is what the game is worth. Reprising our formulas from pages 97 and 102 in Chapter 8:

> The *expected value* of a game is the average of all the payoffs weighted according to their probability.

We'll work out lots of examples below. Before we do that, though, we want to introduce a second quantity associated to a game G. This will measure how much, on average, the actual payoff differs from the expected value; it's called the *variance* of the game, and denoted var(G).[1]

As with the expected value, the variance is a weighted average, weighted according to the probabilities of the various possible outcomes. But rather than simply averaging the payoffs, it averages *how much the payoffs differ from the expected value*: for each possible outcome, we take the difference between the payoff associated to that outcome and the expected value of the game, square it, and take the weighted average of these. To express this as a formula, if we denote the expected value of the game simply by ev, we define

$$\text{var}(G) = p_1(a_1 - \text{ev})^2 + p_2(a_2 - \text{ev})^2 + \cdots + p_k(a_k - \text{ev})^2.$$

> The *variance* of a game is the average of the square of the differences between the payoffs and the expected value, weighted according to their probability:
>
> $$\text{var}(G) = p_1(a_1 - \text{ev})^2 + p_2(a_2 - \text{ev})^2 + \cdots + p_k(a_k - \text{ev})^2.$$

As the name suggests, the variance of a game measures how much the actual payoffs tend to deviate from the expected value. By way of an example, consider two possible games. In the first, we simply flip a fair coin, with outcomes H and T; we pay off \$2 for a head, and nothing for a tail. In tabular form,

outcomes	H	T
probability	1/2	1/2
payoff	2	0

---

[1] In many texts on statistics, the variance of a game G is denoted $\sigma^2(G)$.

In this game, half the time we'll win \$2, and half the time nothing; the following calculation reveals that the expected value, accordingly, is one dollar:

$$\text{ev} = \frac{1}{2} \cdot 2 + \frac{1}{2} \cdot 0 = 1.$$

By way of contrast, consider a second game: this time, we roll a single die, with a payoff of \$6 if it comes up 6 and nothing otherwise. We can also view this as a two-outcome game, with table

outcomes	6	not 6
probability	1/6	5/6
payoff	6	0

This game also has an expected value of \$1: one-sixth of the time we win \$6, and the rest of the time nothing, so

$$\text{ev} = \frac{1}{6} \cdot 6 + \frac{5}{6} \cdot 0 = 1.$$

In other words, both games have expected value of 1. But now let's compute and compare the variances. In the first game, half the time we win \$2 and half the time \$0. In either case, the difference between the actual payoff and the expected value is 1, and so the variance should be the square of that quantity, or 1 again: written out, we have

$$\text{var} = \frac{1}{2} \cdot (2 - 1)^2 + \frac{1}{2} \cdot (0 - 1)^2 = \frac{1}{2} + \frac{1}{2} = 1.$$

In the second game, five-sixths of the time our payoff (0) differs from the expected value (1) by 1, and the other one-sixth of the time it differs by 5. So the variance is

$$\begin{aligned}
\text{var} &= \frac{1}{6} \cdot (6 - 1)^2 + \frac{5}{6} \cdot (0 - 1)^2 \\
&= \frac{1}{6} \cdot 25 + \frac{5}{6} \cdot 1 \\
&= \frac{25}{6} + \frac{5}{6} \\
&= 5.
\end{aligned}$$

Again, this larger value of the variance reflects the fact that the payoffs in the second game vary more.

Now that we have worked out the expected value and variance for a single flip of a fair coin, the case where the coin is unfair—where it comes up heads with probability p, and tails with probability $1 - p$—is similar. If we assign the payoff 1 to heads and 0 to tails, the game looks like:

outcomes	heads	tails
probability	p	$1 - p$
payoff	1	0

The expected value is p, naturally enough; as for the variance, we just plug the values $p_1 = p$, $p_2 = 1 - p$, $a_1 = 1$, $a_2 = 0$, and ev $= p$ into the formula for variance (sorry for the algebra):

$$\begin{aligned}
\text{var} &= p_1(a_1 - \text{ev})^2 + p_2(a_2 - \text{ev})^2 \\
&= p \cdot (1 - p)^2 + (1 - p) \cdot (0 - p)^2 \\
&= p \cdot (1 - p)^2 + (1 - p) \cdot p^2 \\
&= p \cdot (1 - p) \cdot (1 - p + p) \\
&= p \cdot (1 - p).
\end{aligned}$$

We'll use this later on, when we talk about multiple iterations of games.

Exercise 12.1.1. Suppose you play a game in which you simply flip one (fair) coin; if it comes up heads, you win \$10, and if it comes up tails you have to pay \$8. What is the expected value of this game, and what is its variance?

Exercise 12.1.2. In a game we'll call G, one person is chosen at random from a high school class consisting of six 18-year-olds, three 17-year-olds, and three 19-year-olds; the payoff of the game is the age of the person chosen. A second game, H, is similar, but now a person is chosen at random from a family whose members have ages 2, 3, 5, 7, 11, 13, 17, 19, 41, 43, 57, 83, and 97; again, the payoff of the game is the age of the person chosen. Without calculating anything, say which game has bigger variance.

Exercise 12.1.3. Suppose you play a game in which you roll one die, and the payoff is the number of dots showing. What is the expected value and what is the variance of this game?

Exercise 12.1.4. Suppose you play a game in which you are dealt one card from a standard deck; you win \$10 if the card is an ace, and lose \$1 otherwise. What is the approximate variance of this game?

12.2 CHANGING THE PAYOFFS

At this point, we want to consider various ways of altering a game, and what effect this may have on things like its expected value and variance. In this section, we consider only very simple alterations that leave the underlying outcomes and probabilities the same but change the payoffs.

For example, suppose you have a game G, and you change it by adding a fixed amount—let's call it c—to each payoff (including payoffs that are 0). In other words, the event is the same; the possible outcomes P_i are the same; the probability p_i of each possible outcome is the same; but if a_i is the payoff for outcome P_i in the original game, in the new game it's $a_i + c$. This is really not much of a change in the game—if c is positive, we're just that much better off, whatever the outcome, and if c is negative we're that much worse off. We'll give the new game a different name, though; we'll call it $G + c$.

What is the expected value of $G + c$? Pretty much what you'd think: since the payoff has changed by c no matter what the outcome, the expected value should be just c more than the expected value of the original game. The formula bears this out: the new game has expected value

$$\text{ev}(G + c) = p_1(a_1 + c) + p_2(a_2 + c) + \cdots + p_k(a_k + c)$$
$$= p_1 a_1 + p_2 a_2 + \cdots + p_k a_k + (p_1 + p_2 + \cdots + p_k)c$$
$$= \text{ev}(G) + c,$$

since the sum of all the p_is is 1.

What about the variance? Well, if the payoffs all go up by c, and the expected value goes up by c, the *difference* between each payoff and the expected value of the game doesn't change at all! The formula for the variance of $G + c$ reads exactly the same as the formula for $\text{var}(G)$; so

$$\text{var}(G + c) = \text{var}(G).$$

To sum up: if we add a fixed number c to the payoffs of a game G, the new game $G + c$ has the same variance as G, and its expected value is c greater.

Let us now consider a similar operation to the last. As before, the outcomes and probabilities of a game G are going to stay the same, and we're just going to change the payoffs. This time, though, instead of adding a fixed number to each payoff, we're going to multiply them all by a fixed number d. We'll call the new game dG; it has the same outcomes P_i and probabilities p_i as G, but the payoff a_i is now da_i.

Again, if we know the expected value and variance of the original game, we can figure them out for the new one, though it's maybe slightly less obvious. Start with the expected value: if every payoff is multiplied by d, then when we apply the formula for expected value to dG we have

$$\text{ev}(dG) = p_1(da_1) + p_2(da_2) + \cdots + p_k(da_k)$$
$$= d\left(p_1 a_1 + p_2 a_2 + \cdots + p_k a_k\right)$$
$$= d \cdot \text{ev}(G),$$

so in other words the expected value gets multiplied by d.

As for the variance, think about it: if the actual payoff gets multiplied by d, and the expected value gets multiplied by d, then the *difference between them is also multiplied by d*. Since the variance is a weighted average of the squares of these differences, then, it should get multiplied by d^2. And it checks out: if the expected value of the old game is ev, the expected value of the new one is $d \cdot$ ev, and its variance is

$$\text{var}(dG) = p_1(da_1 - d \cdot \text{ev})^2 + p_2(da_2 - d \cdot \text{ev})^2 + \cdots + p_k(da_k - d \cdot \text{ev})^2$$
$$= d^2\left(p_1(a_1 - \text{ev})^2 + p_2(a_2 - \text{ev})^2 + \cdots + p_k(a_k - \text{ev})^2\right)$$
$$= d^2 \cdot \text{var}(G).$$

In other words, when we multiply the payoffs of a game by a number d, we multiply its expected value by d and its variance by d^2.

Exercise 12.2.1.

1. By adding a constant to the payoffs of a game G, can it always be arranged so that the expected value of the game is zero?
2. By adding a constant to the payoffs of a game G, can it always be arranged so that the variance of the game is one?
3. By multiplying the payoffs of a game G by a constant, can it always be arranged so that the expected value of the game is zero?
4. By multiplying the payoffs of a game G by a constant, can it always be arranged so that the variance of the game is one?

12.3 THE NORMALIZED FORM OF A GAME

When we want to compare or combine two games, it helps if they have the same expected value and variance. Happily, we can convert any game into a game with expected value 0 and variance 1, just by applying the operations of adding a number and multiplying by a number. The resulting game is called the *normalized form* of the original.

Here's how it works: suppose you start with any game G. Say the expected value of G is ev. To start, we simply subtract the number ev from each payoff; as we saw, the new game $G - \text{ev}$ will have expected value 0. We also saw that the new game $G - \text{ev}$ will have variance var the same as the original game G. To make that 1, we just divide by the square root of var. The expected value of $G - \text{ev}$ is 0, so that doesn't change when we multiply it by $\sqrt{\text{var}}$; we conclude that the new game, which we can write as

$$G_0 = \frac{G - \text{ev}}{\sqrt{\text{var}}},$$

has expected value 0 and variance 1.

For any game G, its *normalized form* is the game

$$G_0 = \frac{G - \text{ev}}{\sqrt{\text{var}}},$$

which has the same outcomes and probabilities, but with payoffs adjusted so that the expected value is 0 and the variance is 1.

For example, suppose we flip a fair coin, with payoff 1 if it comes up heads and 0 if it's tails. As we saw, the tabular form of this game is

outcomes	H	T
probability	$1/2$	$1/2$
payoff	1	0

This game G has expected value $1/2$, and variance

$$\text{var} = \frac{1}{2}\left(1 - \frac{1}{2}\right)^2 + \frac{1}{2}\left(0 - \frac{1}{2}\right)^2 = \frac{1}{4}.$$

The normalized form of this simple game would be

$$G_0 = \frac{G - \frac{1}{2}}{\frac{1}{2}} = 2\left(G - \frac{1}{2}\right).$$

In other words, in the normalized version G_0, the payoffs would be 1 for heads and -1 for tails.

Problem 12.3.1. Find the normalized form of the game in which you flip an unfair coin that comes up heads p of the time and tails $1 - p$ of the time, with payoff $a_H = 1$ for heads and $a_T = 0$ for tails.

Solution. To start, the tabular form of this game is like the one we just considered, with different probabilities:

outcomes	H	T
probability	p	$1 - p$
payoff	1	0

As we've seen, the expected value of this game is $\text{ev}(G) = p \cdot 1 + (1 - p) \cdot 0 = p$; the variance, accordingly, is

$$\begin{aligned}
\text{var}(G) &= p \cdot (1 - p)^2 + (1 - p) \cdot (0 - p)^2 \\
&= p(1 - 2p + p^2) + (1 - p)p^2 \\
&= p - p^2.
\end{aligned}$$

Thus, the normalized form of the game G would be the same game, but with payoffs

$$\frac{a_H - \text{ev}}{\sqrt{\text{var}}} = \frac{1 - p}{\sqrt{p - p^2}} \quad \text{and} \quad \frac{a_T - \text{ev}}{\sqrt{\text{var}}} = \frac{-p}{\sqrt{p - p^2}};$$

or, in tabular form,

outcomes	H	T
probability	p	$1 - p$
payoff	$\dfrac{1-p}{\sqrt{p-p^2}}$	$\dfrac{-p}{\sqrt{p-p^2}}$

$\square$

One more bit of terminology/notation. The square root of the variance appeared in the expression above for the normalized form of a game. In fact it appears quite often—often enough that its has its own name and symbol: we call the quantity $\sqrt{\operatorname{var}(G)}$ the *standard deviation* of the game G, and we'll denote it by $\operatorname{std}(G)$.[2]

12.4   ADDING GAMES

A final operation on games is trickier but also more interesting than the first two considered in Section 12.2. To illustrate it, we'll start with an example. The basic idea is, we want to combine two games by playing them both and adding up the payoffs. For example, consider the following two games. In the first game, G, you flip one fair coin; if the coin comes up heads, you win 1 and if it comes up tails you win nothing. In the second game, which we'll call H, you roll one fair die; if the die comes up a 6, you win 3 and if the die comes up anything but a 6, you get 0. We've seen that the first game has expected value and variance

$$\operatorname{ev}(G) = \frac{1}{2} \quad \text{and} \quad \operatorname{var}(G) = \frac{1}{4}.$$

As for the second, since you win 3 one-sixth of the time, H has expected value $1/2$; as for the variance, we plug in and find

$$\begin{aligned}
\operatorname{var}(H) &= \frac{1}{6}\left(3 - \frac{1}{2}\right)^2 + \frac{5}{6}\left(0 - \frac{1}{2}\right)^2 \\
&= \frac{1}{6}\left(\frac{5}{2}\right)^2 + \frac{5}{6}\left(\frac{1}{2}\right)^2 \\
&= \frac{25 + 5}{24} = \frac{5}{4}.
\end{aligned}$$

(We could also have gotten this by observing this is just one-half the game we discussed in the first section on variance; that was the same, but with a payoff of 6, and its variance was 5.)

We can combine these simply by both flipping a coin *and* rolling a die; we take the payoffs from each and add them up. The combined game, which we'll call $G + H$, has four possible outcomes: heads/6, heads/not 6, tails/6, and tails/not 6. We know the payoffs for each: for example, for the coin comes up heads and the die a 6, we win 1 for the coin and 3 for the die, for a total of 4; if the coin comes up heads and the die a "not 6," we just win the 1, and so on. Here's the complete list of payoffs:

outcomes	heads/6	heads/not 6	tails/6	tails/not 6
payoff	4	1	3	0

We also know how to calculate the probabilities: assuming the two events are independent, the probability of heads/6 would be $\frac{1}{2} \cdot \frac{1}{6} = \frac{1}{12}$; the probability of heads/not 6 would be $\frac{1}{2} \cdot \frac{5}{6} = \frac{5}{12}$, and so on. Here's the whole table for $G + H$:

---

[2]   As we indicated, most texts on statistics denote the variance of a game G by the symbol $\sigma^2(G)$; naturally enough, in those texts the standard deviation of G is denoted simply $\sigma(G)$.

outcomes	heads/6	heads/not 6	tails/6	tails/not 6
probability	1/12	5/12	1/12	5/12
payoff	4	1	3	0

From this, we can get the expected value and the variance: the first is

$$\text{ev}(G+H) = \frac{1}{12} \cdot 4 + \frac{5}{12} \cdot 1 + \frac{1}{12} \cdot 3 + \frac{5}{12} \cdot 0$$
$$= \frac{12}{12} = 1.$$

So the first thing to notice is that the expected value of the sum $G+H$ is the sum of the expected values of G and H. This sort of makes sense: on average, you'll come away from Game G one-half a unit richer than when you went in, and likewise with H; if you play them both, then, on average you'll come away 1 unit richer.

The variance is less clear. Let's calculate it:

$$\text{var}(G+H) = \frac{1}{12} \cdot (4-1)^2 \frac{5}{12} \cdot (1-1)^2 \frac{1}{12} \cdot (3-1)^2 \frac{5}{12} \cdot (0-1)^2$$
$$= \frac{18}{12} = \frac{3}{2}.$$

Remarkably, this is likewise the sum of the variances of G and H! This isn't something that should be obvious to you, but it's true: the variance of a sum of games is the sum of their variances. We'll see why this is true below in the special case of two two-outcome games; for two arbitrary games, you'll just have to take our word for it.

There's a crucial hypothesis that made these calculations work out the way that they did: *that the games G and H are independent.* If the outcome of G affected the outcome of H then we would no longer expect the expected value and variance of the combined games to be computed in this way. When we speak about "adding games" in the future, we will always assume that the games being added are independent in this way.

To sum up what we've seen:

- If you add a fixed number c to a game, the expected value increases by c and the variance doesn't change.
- If you multiply a game by a number d, the expected value is multiplied by d and the variance is multiplied by d^2.
- If you add two independent games, the expected value of the sum is the sum of the expected values, and the variance of the sum is the sum of the variances.

In tabular form (we seem to be doing a lot of tables recently):

operation	expected value	variance
adding c	$\text{ev}(G+c) = \text{ev}(G) + c$	$\text{var}(G+c) = \text{var}(G)$
multiplying by d	$\text{ev}(dG) = d \cdot \text{ev}(G)$	$\text{var}(dG) = d^2 \cdot \text{var}(G)$
adding games	$\text{ev}(G+H) = \text{ev}(G) + \text{ev}(H)$	$\text{var}(G+H) = \text{var}(G) + \text{var}(H)$

To justify the formula for the variance of the sum of independent games, which we've claimed is just the sum of the variances, let us consider carefully what happens when we add two games. To more clearly communicate the idea of what's going on without bogging ourselves down in proliferating indices, we'll just work with the relatively simple case of two-outcome games. So: let G be a game with two outcomes P_1 and P_2, with probabilities p_1 and p_2 and payoffs a_1 and a_2; let H likewise be a game with outcomes Q_1 and Q_2, with probabilities q_1 and q_2 and payoffs b_1 and b_2. If we play both games, there are four possible outcomes; here's a table with their payoffs:

outcomes	P_1 and Q_1	P_1 and Q_2	P_2 and Q_1	P_2 and Q_2
payoff	$a_1 + b_1$	$a_1 + b_2$	$a_2 + b_1$	$a_2 + b_2$

Now, we're assuming the two games are independent, so the probability of any particular combination—for example, P_1 and Q_2—is just the product of the probabilities of P_1 and Q_2 individually. We can thus fill in the remaining line of the table:

outcomes	P_1 and Q_1	P_1 and Q_2	P_2 and Q_1	P_2 and Q_2
probability	$p_1 q_1$	$p_1 q_2$	$p_2 q_1$	$p_2 q_2$
payoff	$a_1 + b_1$	$a_1 + b_2$	$a_2 + b_1$	$a_2 + b_2$

As a warm-up for the variance, let's calculate the expected value: this is

$$\text{ev}(G + H) = p_1 q_1(a_1 + b_1) + p_1 q_2(a_1 + b_2) + p_2 q_1(a_2 + b_1) + p_2 q_2(a_2 + b_2).$$

This may look like a mess, but if we just multiply everything out and regroup the terms—remembering that p_1 and p_2 have to add up to 1, and likewise for q_1 and q_2—we see that this is

$$\begin{aligned}
\text{ev}(G + H) &= p_1 q_1(a_1 + b_1) + p_1 q_2(a_1 + b_2) + p_2 q_1(a_2 + b_1) + p_2 q_2(a_2 + b_2) \\
&= p_1 q_1 a_1 + p_1 q_1 b_1 + p_1 q_2 a_1 + p_1 q_2 b_2 \\
&\quad + p_2 q_1 a_2 + p_2 q_1 b_1 + p_2 q_2 a_2 + p_2 q_2 b_2 \\
&= p_1 a_1(q_1 + q_2) + p_2 a_2(q_1 + q_2) + q_1 b_1(p_1 + p_2) + q_2 b_2(p_1 + p_2) \\
&= p_1 a_1 + p_2 a_2 + q_1 b_1 + q_2 b_2 \\
&= \text{ev}(G) + \text{ev}(H).
\end{aligned}$$

The calculation for the variance is similar, though, as you might expect, more complicated. (Aren't you glad we're only doing two-outcome games here?) The one thing we can do beforehand to clean it up a little is to shift the games G and H so they have expected value 0—that is, subtract $\text{ev}(G)$ from G and $\text{ev}(H)$ from H. This doesn't affect the variance of either G or H; and, since we're just subtracting the sum $\text{ev}(G) + \text{ev}(H)$ from $G + H$, it doesn't affect the variance of $G + H$ either. So we can just assume that the expected value of G and H are both 0, and (by what we just did) that the expected value of $G + H$ is 0 as well. This simplifies the formula for variance somewhat: we have

$$\begin{aligned}
\text{var}(G + H) = {}& p_1 q_1(a_1 + b_1)^2 + p_1 q_2(a_1 + b_2)^2 \\
&+ p_2 q_1(a_2 + b_1)^2 + p_2 q_2(a_2 + b_2)^2
\end{aligned}$$

$$\begin{aligned}
&= p_1 q_1 a_1^2 + 2 p_1 q_1 a_1 b_1 + p_1 q_1 b_1^2 + p_1 q_2 a_1^2 + 2 p_1 q_2 a_1 b_2 + p_1 q_2 b_2^2 \\
&\quad + p_2 q_1 a_2^2 + 2 p_2 q_1 a_2 b_1 + p_2 q_1 b_1^2 + p_2 q_2 a_2^2 + 2 p_2 q_2 a_2 b_2 + p_2 q_2 b_2^2 \\
&= p_1 a_1^2 + p_2 a_2^2 + q_1 b_1^2 + q_2 b_2^2 + 2 \left(p_1 a_1 + p_2 a_2 \right) \left(q_1 b_1 + q_2 b_2 \right) \\
&= \mathrm{var}(G) + \mathrm{var}(H),
\end{aligned}$$

since the expected values $\mathrm{ev}(G) = p_1 a_1 + p_2 a_2$ and $\mathrm{ev}(H) = q_1 b_1 + q_2 b_2$ are both 0.

By way of illustration, let's take the simple-minded game G in the last example above—where we flip a single fair coin, and the payoff for heads is 1—and see what happens if we play the game repeatedly. The first example would be $G + G$; that is, the game where we flip two coins, and pay off 1 for each head. We can describe this as a game with three possible outcomes—no heads, one head, or two heads—and the table looks like this:

outcomes	no heads	one head	two heads
probability	1/4	1/2	1/4
payoff	0	1	2

Similarly, we could consider the game $G + G + G$, where we flip three coins and again pay off the number of heads; this has table

outcomes	no heads	one head	two heads	three heads
probability	1/8	3/8	3/8	1/8
payoff	0	1	2	3

Now we're finally returning to the questions asked at the beginning of this chapter, but we still don't have enough computational tools to answer them. How can we estimate the *distribution* of outcomes when a game G is iterated not just a handful of times, but 100 or 1,000 or 1,000,000 times over? We'll learn the answer in the next chapter.

Exercise 12.4.1. Verify directly, from the above tables, that the variances of $G + G$ and $G + G + G$ are
$$\mathrm{var}(G + G) = \frac{1}{2} \quad \text{and} \quad \mathrm{var}(G + G + G) = \frac{3}{4}.$$

Exercise 12.4.2. How does the game $G + G$ (the same game played twice) relate to the game $2 \cdot G$ (the same game with payoffs doubled)?

Exercise 12.4.3.

1. Suppose you play a game in which you roll one die, and the payoff is the number of dots showing. What is the expected value and the variance of this game?
2. What is the normalized form of the game just described?
3. Now suppose you play a game in which you roll 100 dice, and the payoff is the total number of dots showing on all the dice. What is the expected value and the variance of this game?

13 The normal distribution

When a game is so addictive that we play it repeatedly, more times than anyone would like to count, it no longer makes sense to compute the exact probabilities of any particular outcomes. Instead, a better question is what is the overall distribution of outcomes of many iterates of the game? Remarkably, if the game is played enough times, then these distributions all have the same overall shape *no matter what the probabilities and payoffs of the original game were*! In this chapter we introduce the remarkable *normal distribution* of a game and use it to answer the questions posed at the start of Chapter 12.

13.1 GRAPHICAL REPRESENTATION OF A GAME

The tables in the last chapter certainly convey all the information we need to know about the games they represent. But they're just a bunch of numbers, and may not give you a qualitative sense of how likely the various payoffs are. A better way to do this is with graphical representations of games: specifically, bar charts. In these charts, we'll arrange the possible payoffs along the horizontal axis, and put a bar over each reflecting the likelihood of that payoff. For example, here's the bar chart representing the game $G + G + G + G$, or in other words flipping 4 fair coins and paying off the number of heads:

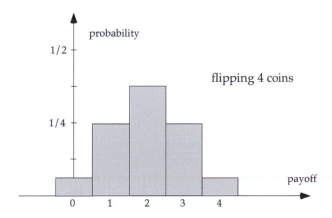

Actually, the notation "$G + G + G + G$" is apt to get awkward if we start repeating games a lot of times, as the title of the last chapter suggests we will. Instead, we'll denote the game consisting of n iterations of the game G by $G(n)$, so that we'd write $G(4)$ instead of $G + G + G + G$. (The reader who is inclined to write "$4 \cdot G$" in place of $G(4)$ is advised to revisit Exercise 12.4.2.)

Note one thing about these bar charts in general: if the width of each bar is 1, the total area of the bars is just the sum of the probabilities associated to the various possible payoffs, and this is just 1.

In any event, here's the bar charts for flipping 6 coins, or $G(6)$:

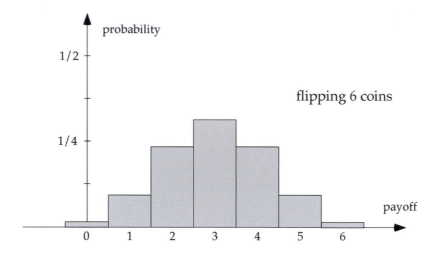

and for $G(8)$:

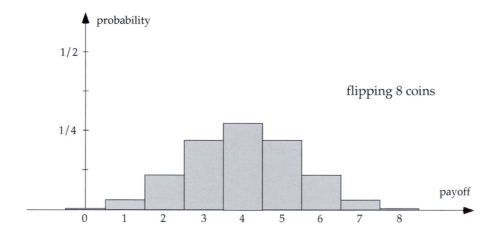

Looking at these charts, we see several patterns: they're all symmetrical around the middle payoff, which is $n/2$ if the game is $G(n)$; the likelihood of the payoff k increases as k goes from 0 to $n/2$, and then decreases the rest of the way.

If we try to compare the charts directly, though, we have a number of problems. For one, the midpoint of the charts keeps moving to the right as n gets bigger. For another, as the whole bar chart gets more and more spread out, it also gets flatter and flatter: if we flip 10 coins, the most likely payoff is 5, which will occur

$$\frac{\binom{10}{5}}{2^{10}} \sim 0.246,$$

or a little less than one-quarter of the time. If we flip 20 coins, the most likely payoff is 10, which occurs

$$\frac{\binom{20}{10}}{2^{20}} \sim 0.176;$$

and the likelihood of exactly 25 heads in 50 flips is

$$\frac{\binom{50}{25}}{2^{50}} \sim 0.112;$$

the likelihood of exactly 50 heads in 100 flips is only 0.080, and the trend continues as n increases. This makes sense: as we said, the total area of the bars in the chart is 1, so as the chart gets more spread out, with more bars, it will simultaneously get flatter. So if we were to ask naively where these bar charts were heading, the answer would be: the x-axis.

But there's a way to fix this, and to make a direct comparison of these games with one another. The first thing to do is to halt the rightward creep, by adjusting the expected value. We know how to do this: just subtract the number $n/2$ from the game; that is, make the payoff for k heads be $k - n/2$ rather than k. Now the payoffs range from $-n/2$ to $n/2$, with 0 the most likely; the bar charts are correspondingly centered around 0. Here's the chart for the adjusted game $G(8) - 4$, for example, where we flip 8 coins and pay off the number of heads minus 4:

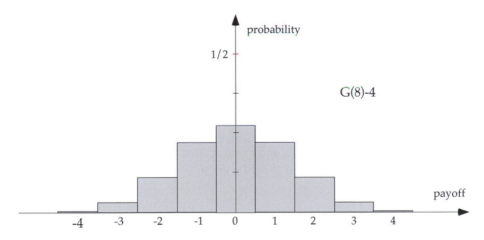

We still have the problem of the charts getting flatter and flatter; but we also have a natural way of dealing with this: we normalize the games so that the variance is 1. Think of it this way: as we flip more and more coins, the possible payoffs spread out further and further away from the center (which is now 0). You'd expect the variance to increase accordingly, and it does: we saw that the variance of our original game G (flipping one coin, and paying off 1 if it comes up heads) is $1/4$, and it follows that the variance of the game $G(n)$ is just n times that: in other words,

$$\text{var}\left(G(n) - \frac{n}{2}\right) = \text{var}(G(n)) = \frac{n}{4}.$$

To make the variance 1, then, we have to divide the payoffs by the square root of this number, or $\sqrt{n}/2$. Basically, as n gets larger, we're shrinking the x-axis on our bar charts by a factor of $\sqrt{n}/2$; the possible payoffs will still spread out, but more of the action will occur close to 0. Note that the game we've arrived at of flipping n coins and paying off

$$a_k = \frac{k - \frac{n}{2}}{\frac{\sqrt{n}}{2}} = \frac{2}{\sqrt{n}}\left(k - \frac{n}{2}\right)$$

is just the normalized form of the game $G(n)$, as described in Section 12.3 above, so we can just call it $G(n)_0$.

We have one more step to go in fixing our bar charts. When we shrink the x-axis by a factor of $\sqrt{n}/2$, the total area of the bars is no longer 1. To recover this—and to keep the bars from flatlining—we need simultaneously to stretch the vertical direction by the same factor. The resulting bar chart—basically, the bar chart of the game $G(n)_0$, but with the vertical axis stretched by $\sqrt{n}/2$—we'll call the *normalized bar chart* of $G(n)$.

And now, having done all this, there's good news: as n increases, *the normalized bar charts of the game $G(n)$ get closer and closer to the area under a single, fixed curve*—the bell-shaped curve, as it's commonly known. This curve has been known for more than two centuries; it was first described by Gauss, and it is one of the signature achievements of the career of probably the greatest mathematician ever.

The upshot for us is immediate: if we want to answer sort of question posed at the beginning of this chapter—how likely it is that flipping 100 coins will yield 60 or more heads, say—we'd think of it this way: if we were to make a bar chart for the game $G(100)$, the answer to our question would be the total area of the bars to the right of the one at 60. Again, we could in theory calculate this by adding up the heights of all the bars—in other words, adding up the numbers $\binom{100}{k}/2^{100}$ for all k between 60 and 100—but now we have an alternative. We can look at the normalized bar chart of the game $G(100)$, and take the normal distribution as an good approximation to it. The total area of the bars to the right of the one at 60 is then approximated by the area under the corresponding portion of the bell-shaped curve, and we can look this up online or in a table.

Needless to say, the bell-shaped curve, called the *normal distribution*, has been extensively studied. It turns out to be the graph of a function, namely

$$f(x) = \frac{1}{\sqrt{2\pi}}e^{-\frac{x^2}{2}}.$$

Now, if you haven't had calculus, you may not be familiar with the number e; it's a naturally occurring constant, approximately equal to 2.7128. Given this, the shape of the graph of this function can be described at least qualitatively: since we're raising the number e, which is bigger than 1, to a negative power $(-x^2/2)$, the result will be largest when the exponent is smallest, which will be the case when $x = 0$; as x moves away from 0, x^2 gets larger, and $e^{-x^2/2}$ gets correspondingly smaller.

But we don't need to know any of this! Even if we were wizards at calculus, we still couldn't calculate exactly the area under this curve between two points; we'd have to do what we're going to do anyway, which is as we said look up the numbers in a table, such as the one we've printed here (and also copied to the appendix), created by William Knight:

Probability Content from   -oo to Z

z	0.00	0.01	0.02	0.03	0.04	0.05	0.06	0.07	0.08	0.09
0.0	0.5000	0.5040	0.5080	0.5120	0.5160	0.5199	0.5239	0.5279	0.5319	0.5359
0.1	0.5398	0.5438	0.5478	0.5517	0.5557	0.5596	0.5636	0.5675	0.5714	0.5753
0.2	0.5793	0.5832	0.5871	0.5910	0.5948	0.5987	0.6026	0.6064	0.6103	0.6141
0.3	0.6179	0.6217	0.6255	0.6293	0.6331	0.6368	0.6406	0.6443	0.6480	0.6517
0.4	0.6554	0.6591	0.6628	0.6664	0.6700	0.6736	0.6772	0.6808	0.6844	0.6879
0.5	0.6915	0.6950	0.6985	0.7019	0.7054	0.7088	0.7123	0.7157	0.7190	0.7224
0.6	0.7257	0.7291	0.7324	0.7357	0.7389	0.7422	0.7454	0.7486	0.7517	0.7549
0.7	0.7580	0.7611	0.7642	0.7673	0.7704	0.7734	0.7764	0.7794	0.7823	0.7852
0.8	0.7881	0.7910	0.7939	0.7967	0.7995	0.8023	0.8051	0.8078	0.8106	0.8133
0.9	0.8159	0.8186	0.8212	0.8238	0.8264	0.8289	0.8315	0.8340	0.8365	0.8389
1.0	0.8413	0.8438	0.8461	0.8485	0.8508	0.8531	0.8554	0.8577	0.8599	0.8621
1.1	0.8643	0.8665	0.8686	0.8708	0.8729	0.8749	0.8770	0.8790	0.8810	0.8830
1.2	0.8849	0.8869	0.8888	0.8907	0.8925	0.8944	0.8962	0.8980	0.8997	0.9015
1.3	0.9032	0.9049	0.9066	0.9082	0.9099	0.9115	0.9131	0.9147	0.9162	0.9177
1.4	0.9192	0.9207	0.9222	0.9236	0.9251	0.9265	0.9279	0.9292	0.9306	0.9319
1.5	0.9332	0.9345	0.9357	0.9370	0.9382	0.9394	0.9406	0.9418	0.9429	0.9441
1.6	0.9452	0.9463	0.9474	0.9484	0.9495	0.9505	0.9515	0.9525	0.9535	0.9545
1.7	0.9554	0.9564	0.9573	0.9582	0.9591	0.9599	0.9608	0.9616	0.9625	0.9633
1.8	0.9641	0.9649	0.9656	0.9664	0.9671	0.9678	0.9686	0.9693	0.9699	0.9706
1.9	0.9713	0.9719	0.9726	0.9732	0.9738	0.9744	0.9750	0.9756	0.9761	0.9767
2.0	0.9772	0.9778	0.9783	0.9788	0.9793	0.9798	0.9803	0.9808	0.9812	0.9817
2.1	0.9821	0.9826	0.9830	0.9834	0.9838	0.9842	0.9846	0.9850	0.9854	0.9857
2.2	0.9861	0.9864	0.9868	0.9871	0.9875	0.9878	0.9881	0.9884	0.9887	0.9890
2.3	0.9893	0.9896	0.9898	0.9901	0.9904	0.9906	0.9909	0.9911	0.9913	0.9916
2.4	0.9918	0.9920	0.9922	0.9925	0.9927	0.9929	0.9931	0.9932	0.9934	0.9936
2.5	0.9938	0.9940	0.9941	0.9943	0.9945	0.9946	0.9948	0.9949	0.9951	0.9952
2.6	0.9953	0.9955	0.9956	0.9957	0.9959	0.9960	0.9961	0.9962	0.9963	0.9964
2.7	0.9965	0.9966	0.9967	0.9968	0.9969	0.9970	0.9971	0.9972	0.9973	0.9974
2.8	0.9974	0.9975	0.9976	0.9977	0.9977	0.9978	0.9979	0.9979	0.9980	0.9981
2.9	0.9981	0.9982	0.9982	0.9983	0.9984	0.9984	0.9985	0.9985	0.9986	0.9986
3.0	0.9987	0.9987	0.9987	0.9988	0.9988	0.9989	0.9989	0.9989	0.9990	0.9990

Far Right Tail Probabilities

z	P{Z to oo}	z	P{Z to oo}	z	P{Z to oo}	z	P{Z to oo}
2.0	0.02275	3.0	0.001350	4.0	0.00003167	5.0	2.867 E-7
2.1	0.01786	3.1	0.0009676	4.1	0.00002066	5.5	1.899 E-8
2.2	0.01390	3.2	0.0006871	4.2	0.00001335	6.0	9.866 E-10
2.3	0.01072	3.3	0.0004834	4.3	0.00000854	6.5	4.016 E-11
2.4	0.00820	3.4	0.0003369	4.4	0.000005413	7.0	1.280 E-12
2.5	0.00621	3.5	0.0002326	4.5	0.000003398	7.5	3.191 E-14
2.6	0.004661	3.6	0.0001591	4.6	0.000002112	8.0	6.221 E-16
2.7	0.003467	3.7	0.0001078	4.7	0.000001300	8.5	9.480 E-18
2.8	0.002555	3.8	0.00007235	4.8	7.933 E-7	9.0	1.129 E-19
2.9	0.001866	3.9	0.00004810	4.9	4.792 E-7	9.5	1.049 E-21

The formats of these tables may vary; when we talk about looking up values of the normal distribution in a table, we'll mean this one.

Let's see how this works in practice, for example in the case of the 100 coin flips above. To normalize the game $G(100)$, we start by subtracting its expected value, which is 50; we then divide by the square root of the variance (the variance is $100 \cdot 1/4 = 25$, so the square root of the variance will be 5) to arrive at the normalized form of $G(100)$, which is

$$G(100)_0 = \frac{G(100) - 50}{5}.$$

Now, the payoff 60 in the original game $G(100)$ corresponds to the payoff

$$z = \frac{60 - 50}{5} = 2$$

of the normalized form $G(100)_0$ of $G(100)$. To the extent that the normal distribution approximates the normalized bar chart of $G(100)$, then, the total area of the bars to the right of the one at 60 in the bar chart of $G(100)$ should equal the area under the graph of the normal distribution to the right of 2. And this we can just look up in the tables of the normal distribution that exist in virtually every book on probability. According to any of these tables, the area under the graph to the right of 2 (or, equivalently, the area under the graph to the left of -2) is approximately 0.0228; and this, accordingly, is the (approximate) probability that we get 60 or more heads in 100 flips.

Similarly, if we want to figure out the odds of getting 600 or more heads in 1,000 flips, we can approximate the normalized bar chart of the game $G(1000)$ by the normal distribution. Since the normalized form of the game $G(1000)$ is

$$G(1000)_0 = \frac{G(1000) - 500}{\sqrt{250}} \sim 0.063\big(G(1000) - 500\big),$$

the payoff 600 of the game $G(1000)$ corresponds to the payoff $z = 6.3$ in the normalized game; according to the table, the area under the graph of the normal distribution to the right of 6.3 is on the order of 10^{-10}, or one ten-billionth. Essentially, this will never happen; if you ever flip a coin 1,000 times and come up with 600 or more heads, you can reasonably start looking for another explanation. This game might be rigged.

Let's do some more examples:

Problem 13.1.1. Suppose you flip a fair coin 400 times. What are the odds that you'll get more than 210 heads? More than 220? More than 230?

Solution. We're talking about the game $G(400)$ here, and the first thing is to find its normalized form. This is straightforward: the expected value is 200, and the variance is $400 \cdot 1/4$, or 100. The normalized form of the game is thus

$$G(400)_0 = \frac{G(400) - 200}{10}.$$

The condition that the payoff in $G(400)$ is more than 210 is equivalent to the condition that the payoff in the normalized game $G(400)_0$ is more than

$$z = \frac{210 - 200}{10} = 1;$$

and this is approximately the area to the right of the line $z = 1$ under the bell-shaped curve. If we look up the value $z = 1$ in the table, it tells us the area to the left of $z = 1$ under the curve is 0.8413; the area to the right is then $1 - 0.8413 = 0.1587$. The likelihood of getting 210 or more heads is thus roughly 0.16, or slightly less than 1 in 6.

The other parts of the problem are done similarly: to say that the payoff in the game $G(400)$ is bigger than 220 is to say the payoff in the normalized game $G(400)_0$ is greater than

$$\frac{220 - 200}{10} = 2;$$

according to the table, the area under the bell-shaped curve to the left of the line $z = 2$ is 0.9772, and the area to the right correspondingly is $1 - 0.9772 = 0.0228$; so, the probability of more than 220 heads in 400 flips is about 2.3%, or about 1 in 44. And as for the possibility of getting 230 or more heads, this corresponds to the value

$$\frac{230 - 200}{10} = 3$$

of the normalized game; according to the tables, the area to the right of the line $z = 3$ is $1 - 0.9987 = 0.0013$, so the probability of this happening is about 1 in 750. $\square$

Problem 13.1.2. Now let's flip a coin 1,000 times, and ask:

1. What is the likelihood of getting between 480 and 520 heads?
2. How about between 470 and 520?

Solution. Again, we start by writing down the normalized form of the game $G(1000)$; since the variance is

$$\text{var}\big(G(1000)\big) = 1000 \cdot \frac{1}{4} = 250,$$

this is

$$G(1000)_0 = \frac{G(1000) - 500}{\sqrt{250}} \sim \frac{G(1000) - 500}{15.81}.$$

The payoffs 480 and 520 of the original game $G(1000)$ correspond to the payoffs

$$\frac{480 - 500}{15.81} = -1.265 \quad \text{and} \quad \frac{520 - 500}{15.81} = 1.265,$$

so the answer to the first question should be approximately the area of the graph of the bell-shaped curve between $z = -1.265$ and $z = 1.265$.

How do we get this from the table? Well, the table says that the area to the left of the line $z = 1.265$ is 0.8971 (we're taking the value halfway between the value 0.8962 for $z = 1.26$ and the value 0.8990 for $z = 1.27$); so the area to the right of this line is

$$1 - 0.8971 = 0.1029.$$

By symmetry, the area to the left of the line $z = -1.265$ is the same, and so the area in the middle is

$$1 - 0.1029 - 0.1029 = 0.7942;$$

so the number of heads will be between 480 and 520 almost four-fifths of the time.

The second problem we do similarly. The payoff 470 of the original game $G(1000)$ corresponds to the payoff

$$\frac{470 - 500}{15.81} = -1.90.$$

According to the table, the area to the left of the line $z = 1.90$ is 0.9713. By symmetry, this is the same as the area to the right of the line $z = -1.90$, and so the area to the left of the line $z = -1.90$ is $1 - 0.9713 = 0.0287$. The area in between the lines $z = -1.90$ and $z = 1.265$ is then

$$1 - 0.1029 - 0.0287 = 0.865,$$

or almost 7 times in 8. $\square$

We should end this section with a little note about boundary values. The astute reader will have noticed that, in the last problem—"if we flip 1,000 coins, what is the likelihood of getting between 470 and 520 heads?"—we didn't specify if we meant "between 470 and 520" to be inclusive or not: does getting exactly 470 heads count? What's more, the method of finding an (approximate) solution doesn't seem to take this distinction into account: we just find the values $z = -1.90$ and $z = 1.265$ of the normalized form of the game corresponding to 470 and 520, and look up the area under the bell-shaped curve between these values. What's up with this?

The answer is, we're really only getting an approximate answer; and the likelihood of getting exactly 480 heads—a value right on the boundary of the range we're interested in—is very small, and within the error margin of our approximation. Specifically, the odds of getting exactly 480 heads is

$$\frac{\binom{1,000}{480}}{2^{1,000}} \sim 0.0042,$$

or less than half a percent. If we meant "inclusive" in the problem, and we wanted to maximize the accuracy of our approximation, we could say instead "between 469.5 and 520.5," so as to remove the ambiguity. Likewise, if we meant exclusive—so that 470 and 520 didn't count—we could say "between 470.5 and 519.5." In the context of this course, though, we won't worry about these finer points.

Exercise 13.1.3. Let G be the game in which you flip an unfair coin, one that comes up heads on average 60% of the time and tails 40% of the time; the payoff is 1 if the coin comes up heads and 0 if it's tails. What is the normalized form of the game $G(100)$?

13.2 EVERY GAME IS NORMAL

Now, all the stuff we've been doing in the last section is terrific if you're really, really into coin flipping. But let's face it: the normal distribution would seem to have pretty limited scope, if all it can do is estimate the odds that the number of heads in a bunch of coin flips lies in a certain range.

But now comes the amazing part, and the reason why the bell-shaped curve is so ubiquitous in so many areas of human experience. Suppose we start with any game, which we'll call H. This could be any game whatsoever: any (finite) number of possible outcomes, any probabilities, any payoffs. Naturally, the bar chart for H can look like anything you want to make up, as long as the total area of the bars is 1. Now suppose we repeat the game H over and over, in independent iterations, and add up the resulting payoffs; that is, suppose we consider the game

$$H(n) = \underbrace{H + H + \cdots + H}_{n \text{ times}}$$

for large values of n. The remarkable fact is, *no matter what the game H is, for large values of n the normalized bar chart of $H(n)$ will approximate the normal distribution.*

In other words, the normal distribution is not just the "limiting case" for multiple iterations of a coin flip; *it's the limiting case of multiple iterations of every game there is.* So we can use the normal distribution to approximate the probability that the result of a large number of iterations of any game lie in a certain range; and this is what we'll do in the following examples.

We'll start with a simple variation involving dice:

Problem 13.2.1. Suppose you roll 100 dice. What are the odds that 25 or more of them come up 6?

Solution. This is based on a very simple game, which we'll call H: you roll one die, with a payoff of 1 if it comes up 6, and nothing otherwise. In tabular form,

outcomes	6	not 6
probability	1/6	5/6
payoff	1	0

Let's start by figuring out the expected value and variance of H. The expected value we've seen before, and in any case it's easy: if you just win 1 one-sixth of the time, the expected value is 1/6. As for the variance, we just apply the formula from page 161:

$$\text{var}(H) = \frac{1}{6}\left(1 - \frac{1}{6}\right)^2 + \frac{5}{6}\left(0 - \frac{1}{6}\right)^2 = \frac{5}{36}.$$

(But you knew that, because you read and remember Section 12.1, right?)

Now, the question we're dealing with is: in the game $H(100)$, what is the probability of payoff bigger than 25? So the first thing is to find the normalized form of the game $H(100)$. To begin with, we know the expected value of $H(100)$ is just 100 times the expected value of H, or 100/6; the variance is likewise just 100 times the variance of H, or 500/36. So the normalized form of the game is

$$H(100)_0 = \frac{H(100) - \frac{100}{6}}{\sqrt{\frac{500}{36}}} \sim \frac{H(100) - 16.67}{3.727}.$$

The payoff 25 of the game $H(100)$ thus corresponds to the payoff

$$z \sim \frac{25 - 16.67}{3.727} \sim 2.236,$$

and the likelihood of this payoff or higher is approximately the area under the normal distribution to the right of the line $z = 2.236$. According to our table, the area to the left of this line is 0.9873; so the area to the right is

$$1 - 0.9873 = 0.0127,$$

and so this is approximately the probability of getting 25 or more 6s in 100 rolls of a die. □

Let's do another example, this time dealing with a game with more than two possible outcomes:

Problem 13.2.2. Say you roll 100 dice, and add up the numbers showing. On average, you'd expect this sum to be 350. What are the odds it'll be over 400?

Solution. Again, we start by describing the basic game G this is based on: we roll one die, and the payoff is the number showing (so that the game we're playing in the problem is just $G(100)$). Of course, the event in this game is the same as in the game H of the last problem, but the payoffs are different. The table is:

outcomes	1	2	3	4	5	6
probability	1/6	1/6	1/6	1/6	1/6	1/6
payoff	1	2	3	4	5	6

The expected value is

$$\text{ev}(G) = \frac{1}{6} \cdot 1 + \frac{1}{6} \cdot 2 + \frac{1}{6} \cdot 3 + \frac{1}{6} \cdot 4 + \frac{1}{6} \cdot 5 + \frac{1}{6} \cdot 6 = \frac{21}{6},$$

or 3.5; the variance is

$$\begin{aligned}
\text{var}(G) &= \frac{1}{6}(1 - 3.5)^2 + \frac{1}{6}(2 - 3.5)^2 + \frac{1}{6}(3 - 3.5)^2 \\
&\quad + \frac{1}{6}(4 - 3.5)^2 + \frac{1}{6}(5 - 3.5)^2 + \frac{1}{6}(6 - 3.5)^2 \\
&= \frac{1}{6} \cdot \frac{25}{4} + \frac{1}{6} \cdot \frac{9}{4} + \frac{1}{6} \cdot \frac{1}{4} + \frac{1}{6} \cdot \frac{1}{4} + \frac{1}{6} \cdot \frac{9}{4} + \frac{1}{6} \cdot \frac{25}{4} \\
&= \frac{70}{24}.
\end{aligned}$$

The expected value of $G(100)$ is thus 3.5×100, or 350 (as we said in the problem!); the variance is

$$\text{var}(G(100)) = 100 \times \frac{70}{24} \sim 291.7.$$

The normalized form of the game is thus

$$G(100)_0 \sim \frac{G(100) - 350}{\sqrt{291.7}} \sim \frac{G(100) - 350}{17.08}.$$

The value of the normalized game corresponding to a payoff of 400 is then

$$z \sim \frac{400 - 350}{17.08} \sim 2.93,$$

and so the probability of the dice adding up to 400 or more is approximately the area under the bell-shaped curve to the right of the line $z = 2.93$. The area to the left of this line, according to the table, is approximately 0.9983; so the area to its right is 1-0.9983, or 0.0017, and this is our answer. □

Reluctantly breaking a promise we made at the end of Section 8.1, here's an exercise for you to try:

Exercise 13.2.3. Here's a common version of Chuck-A-Luck: you pay \$1 and roll three dice. If you get one 6, you win \$2; if you roll two 6s you win \$3, and if all three come up 6s you win \$4. In tabular form:

outcomes	no 6s	one 6	two 6s	three 6s
probability	$(5/6)^3$	$3\,(1/6)\,(5/6)^2$	$3\,(1/6)^2\,(5/6)$	$(1/6)^3$
what this is, really	0.589	0.347	0.069	0.0046
payoff	-1	1	2	3

1. Find the expected value and variance of this version of Chuck-A-Luck.
2. If you play this version of Chuck-A-Luck 100 times, what is the probability that you'll be ahead—in other words, that the total payoff will be bigger than 0?

Exercise 13.2.4. Suppose you roll 100 dice. What are the odds that 25 or more of them come up 6?

13.3 THE IMPORTANCE OF THE STANDARD DEVIATION

The two examples considered in Section 13.2 deal with the same event—rolling 100 dice—but the payoffs associated with the events differed. In Problem 13.2.1, where we just count the number of 6s rolled, the expected value was $100/6$, or about 16.67. We asked what the likelihood was of our getting 25 or more 6s, which is to say 50% higher than the expected; we found that the probability was 0.0127.

In Problem 13.2.2, by contrast, we asked for the likelihood of the sum of the 100 dice being over 400, or just 14% above the expected value—seemingly, closer to the expected value (as a fraction of the expected value) than in Problem 13.2.1. Yet the probability was much lower: 0.0017, as opposed to 0.0127. What's going on?

The answer is not hard to spot. The point is, suppose we take any game G, and look at a multiple iteration $H = G(n)$ of the game, with n large, so that the normalized bar chart of H looks like the standard bell-shaped curve. If we ask for the likelihood that the outcome of H exceeds a certain value A, what really counts is the corresponding value z of the normalized form H_0 of the game H, that is,

$$z = \frac{A - \mathrm{ev}(H)}{\mathrm{std}(H)}.$$

The point is, we have to consider the difference between A and the expected value $\mathrm{ev}(H)$ of H not as a number, or as a percentage of $\mathrm{ev}(H)$, but *as a multiple of the standard deviation* $\mathrm{std}(H)$ *of* H. For example, if the difference $A - \mathrm{ev}(H)$ is equal to one standard deviation $\mathrm{std}(H)$, the odds of the outcome exceeding A is $1 - 0.8413$, or about 0.1587; if the difference $A - \mathrm{ev}(H)$ is twice the standard deviation $\mathrm{std}(H)$, the odds of the outcome exceeding A is $1 - 0.9772$, or about 0.0228, and so on.

What's more, the standard deviation of the iterate $H = G(n)$ of the game G is proportional to the standard deviation of G (it's just $\sqrt{n} \cdot \mathrm{std}(G)$). So the bigger the variance of the original game G (as a multiple of the expected value), the greater the chance that the outcome of the iterate $G(n)$ will differ from its expected value by a given percentage.

To see how this plays out in practice, note that in the second example the payoffs in the basic game G are spread around, from 1 to 6, whereas in the first example they're all concentrated on one possible outcome with probability $1/6$. The result is that the standard deviation of the basic game G in Problem 13.2.2 is a lower fraction of the expected value than in Problem 13.2.1: in Problem 13.2.1 we have

$$\text{ev}(H) = .1667 \quad \text{and} \quad \text{std}(H) = \sqrt{\frac{5}{36}} = .373,$$

so

$$\frac{\text{std}(H)}{\text{ev}(H)} = 2.236.$$

By contrast, in Problem 13.2.2, we had

$$\text{ev}(G) = 3.5 \quad \text{and} \quad \text{std}(G) = \sqrt{\frac{70}{24}} = 1.708,$$

so

$$\frac{\text{std}(G)}{\text{ev}(G)} = 0.488.$$

The same difference carries over to the iterates $H(100)$ and $G(100)$ considered in the two examples: in Problem 13.2.1 we had

$$\frac{\text{std}(H(100))}{\text{ev}(H(100))} = \frac{3.727}{16.67} = 0.224,$$

while in Problem 13.2.2,

$$\frac{\text{std}(G(100))}{\text{ev}(G(100))} = \frac{17.08}{350} = 0.049.$$

The effect is that the value $z = 2.236$ of the normalized form corresponding to a payoff 50% greater than expected in Problem 13.2.1 is actually *smaller* than the value $z = 2.93$ of the normalized form corresponding to a payoff 14% greater than expected in Problem 13.2.2, and the likelihood of our actual result deviating that much from the expected is correspondingly lower.

The point is this: when we try to figure out how likely it is that the result of a large number of iterations of a game differs from the expected value, *it's not the percentage difference that matters; it's the difference expressed as a multiple of the standard deviation.*

By way of another illustration of this, consider the following series of exercises. These will deal with two games, which we'll call G and H. In both games, the event is simply the selection of a single card from a standard deck of 52. The payoffs differ, though: in the game G, you win $10 if your card is an ace, and lose $1 otherwise; in game H, you win $1 if you get any card from a 9 to an ace (that is, a 9, 10, jack, queen, king, or ace), and lose $1 otherwise. We're going to consider what happens if we play each game 100 times; that is, we're going to look at the games $G(100)$ and $H(100)$. Note that, since the games specify choosing a card at random from a 52-card deck, if we're going to play either game 100 times we have to replace the card selected and reshuffle after each selection (or just use 100 decks of cards; whatever's easier).

Exercise 13.3.1. In the games G and H just introduced:

1. Calculate the expected values of $G(100)$ and $H(100)$. In particular, show that $H(100)$ is more favorable to you over the long run; that is, the expected value of $H(100)$ is greater than the expected value of $G(100)$.
2. Calculate the variances of $G(100)$ and $H(100)$.

Exercise 13.3.2.

1. Using your answer to the previous problem, estimate the likelihood that after playing the game G 100 times, you're ahead (that is, the payoff is positive). Likewise, estimate the probability that you'll be ahead after playing H 100 times.
2. Explain why, if game H is more favorable to you, you have a greater likelihood of being ahead after playing G 100 times.

13.4 POLLING

So far, most our examples have had to do with gambling games. As we've explained before, this is because it's cleaner: in the context of gambling games, we can identify and control all the relevant factors. If we're willing to overlook a lot of potentially significant factors, though (and it is scary how much people seem to be willing to do this), we can apply this sort of estimation in many, many other situations. Here we'll take up one relatively simple example: political polling. In particular, we'll finally be able to answer the pressing question: what do they actually mean when they say that a certain poll has "a 3% margin of error?"

Here's the setup, in the form of an example. Suppose for the moment that we're having a national election for president; our two candidates are none other than Tracy and Paul, who it seems have given up mathematics and aspire to higher office. Say for the moment that 55% of the population favors Tracy (who has learned her lesson from her ill-fated campaign to be class president, and now panders only to majorities), while only 45% favor Paul. We conduct a poll by choosing 1,000 eligible voters at random and asking their preference. The question is, what is the likelihood that the results of our poll will differ from reality by more than, say, 3 percentage points? In other words, what are the odds that, in the responses to our poll, fewer than 52% or more than 58% favor Tracy?

We can totally set this up as a game, and answer that question based on what we've learned in this chapter. To begin with, we introduce a simple game G, in which we choose one random voter and ask their preference; say the payoff is 1 if our random voter is a Tracy supporter and 0 if they're for Paul. The table for this game is

outcomes	Tracy	Paul
probability	.55	.45
payoff	1	0

The expected value of this game is just $.55 \times 1 = .55$, and the variance is accordingly

$$\text{var}(G) = .55(1 - .55)^2 + .45(0 - .55)^2 = 0.2475.$$

The variance of the game $G(1000)$ is then just 1,000 times this, or

$$\text{var}(G(1000)) = 247.5,$$

and the normalized form of the game $G(1000)$ is accordingly

$$G(1000)_0 = \frac{G(1000) - 550}{\sqrt{247.5}} \sim \frac{G(1000) - 550}{15.73}.$$

So, how likely is it that our poll will be off by 3% or more in Tracy's favor—that is, that the actual payoff of our game of $G(1000)$ will be 580 or more? Well, a payoff of 580 or more in $G(1000)$ corresponds to a payoff of

$$z = \frac{580 - 550}{15.73} = \frac{30}{15.73} = 1.907$$

in the normalized game $G(1000)_0$. According to our table, the region under the bell-shaped curve to the left of the line $z = 1.907$ has an area of 0.9717, and the region to the right of that line correspondingly an area of $1 - 0.9717 = 0.0283$. We might then estimate that the likelihood of our poll of 1,000 yielding more than 580 Tracy supporters—given that the actual fraction of the population supporting Tracy is .55—is about 0.0283. The likelihood of our poll being off by 3% or more in the other direction—in other words, that among our 1,000 randomly chosen voters, fewer than 520 will support Tracy—is the same, by symmetry. So, in the end, the likelihood of our poll being off by 3% or more is $0.0283 + 0.0283$, or 0.0566, or a little over 5%.

So: if we know the reality (in this case, 55% of all voters favor Tracy, 45% Paul), we can calculate the likelihood of our poll being off by a given amount. This still doesn't answer our question: what the hell do they mean when they say that a certain poll has "a 3% margin of error?" But at least now we're in a position to answer it. Let's start with our example: by convention, we say a poll that says that 55% of voters prefer Tracy has a 3% margin of error if, assuming that the proportion of voters supporting Tracy is indeed 55%, the likelihood of our poll being off by 3 percentage points or more is under 5%. Since the odds of being off by 3 percentage points in Tracy's favor is roughly the same as the odds of being off that much in Paul's favor, this amounts to saying that the probability of more than 580 of the people surveyed favoring Tracy is under 2.5%, and likewise the probability of the number being below 520 is under 2.5%. If you look at the table, this corresponds roughly to the values $z = \pm2$—in other words, the margin of error is usually taken to be twice the standard deviation of the game.

We do the same in general: we say that a poll showing that $n\%$ of a population feels a certain way "has a margin of error of $\pm k$ percentage points" if, given that the actual proportion of the population feeling that way is $n/100$, the likelihood of our survey yielding a result outside the range of $n \pm k\%$ is under 5%.[1]

Thus, in our example, we can't say that our poll has a margin of error of 3% (assuming that 55% do favor Tracy, the odds of our poll yielding an answer of under 52% or over 58% are just over 5%; but (as you know if you did Exercise 13.4.1 above) if we had sampled 2,000 people we could.

Exercise 13.4.1. Take the same situation as just described, but suppose now we poll 2,000 people rather than 1,000. Now what are the odds our results will differ from reality by more than 3%? How about if we sampled only 500 voters?

Exercise 13.4.2. Again dealing with the same situation as just described, and supposing we poll 1,000 people, what are the odds that our results are off by 5 or more percentage points?

---

[1] You might think that this definition is somewhat backward: it would be better, you might reason, to say that a poll showing 55% of voters preferring Tracy has a 3% margin of error if, assuming that the proportion of voters supporting Tracy is less than 52%, the likelihood of our poll coming up with a figure of 55% or more was under 2.5%, and likewise if the actual percentage was more than 58% the odds of our poll coming up with a figure of 55% or less was under 2.5%. You'd get no argument from us on this one. But in practice, the definition above is commonly adopted because a) it's simpler to state; b) it's easier to calculate; and c) except in extreme circumstances, the two are close enough.

Exercise 13.4.3. Suppose now we conducted a poll of 1,500 voters and found that 60% favored Tracy. Could we say that our poll has a 3% margin of error?

Exercise 13.4.4. Tracy and Paul are running for mayor of Middletown, in which 60% of the voters favor Tracy and 40% support Paul. A poll is conducted in which 100 residents, selected at random, are asked their preference. What is the likelihood that the poll will show a majority in favor of Paul?

13.5   THE FINE PRINT ON POLLING

We need to express a number of caveats here. (Actually, this shouldn't be fine print; it should be the banner headline over any news article reporting poll results.) The first and foremost is, despite the assertion that a given poll "has a 3% margin of error," *there is no guarantee whatsoever that it's not off by more than that.* If you were to flip a fair coin 1,000 times, it's highly unlikely (certainly less than a 5% chance) that you'd get more than 550 or fewer than 450 heads, *but it could happen.* All people mean when they say that a poll has a certain margin of error is that, basically, it's unlikely—specifically, less than a 5% chance—to be off by that much or more.

But that's far from the only caveat. Actual polling is rife with other factors that can skew the results, among them:

- How the sample is chosen: For example, are you limiting yourself to voters with a telephone landline? With an internet connection? Does this affect the result?
- What group are you actually polling: Eligible voters? Registered voters? Likely voters? And do you take people's words for it that they're likely to vote?
- How you phrase the question: Even if your intent is to get honest results, it's all too easy (as people have found) to skew the results of a poll by using (or omitting) key words in the question.

To see how much these other factors affect the reliability of polls, you can read a preliminary analysis of polls taken in the 2010 election at

```
https://fivethirtyeight.com/features/rasmussen-polls-were-
biased-and-inaccurate-quinnipiac-surveyusa-performed-strongly/
```

If you scroll down to the table, you'll see that the polls of every polling firm—even those lauded as having been the most accurate—were off by an average of more than 3%. This is despite the fact that these were almost all polls with a statistical margin of error (that is, the margin of error as we defined it above) of $\pm 3\%$. The point is, the errors didn't arise as a result of random fluctuations; they were a reflection of subtle (and in some cases not-so-subtle) biases of the sort listed here.

13.6   MIXING GAMES

At this point, we've reached the end of what we can reasonably do in this course; to go any further would require more mathematical language and technique than we have. But we should mention two keys facts about the ideas we've discussed so far in this chapter.

It might seem remarkable enough that you can start with any game whatsoever and, if you repeat it enough times and add up the results its normalized bar chart always winds up looking like the bell-shaped curve. But even this may seem limited: after all, how often, in real life, do you play the same game over and over that many times?

In fact, we have some even more amazingly great news: the reality goes even further, and this is a large part of the reason the bell-shaped curve is so ubiquitous.

First, suppose instead of repeating one game a large number of times, we have a whole sequence of possibly different games $G_1, G_2, G_3, \ldots$ Suppose moreover that their expected values and variances all lie in some fixed range. Now say we start adding them up: that is, we take

$$H_1 = G_1,$$
$$H_2 = G_1 + G_2,$$
$$H_3 = G_1 + G_2 + G_3,$$

and so on; in general, let H_n be the sum of the first n games, i.e., $H_n = G_1 + G_2 + \cdots + G_n$. Then, amazingly enough, the same statement holds: *for large enough n, the normalized bar chart of H_n approximates the bell-shaped curve.*

And this, after all, is why the bell-shaped curve is everywhere. Look for example at the distribution of heights within a given population. A person's height is the result of many, many factors: there are many genes that affect height, and in addition there are environmental factors like nutrition that have an effect.

Now, if we want to describe the distribution of heights in the United States, we can think of it as a game H: we pick a person at random in the United States, and the payoff is their height in inches. And this game is, in turn, the sum of many other games: the various genes that contribute to height, family income, culture, diet, and so on. The upshot is that, if we were to create a bar chart displaying this game, it would indeed be shaped very much like the bell-shaped curve.

The same applies to most phenomena that can be thought of as the sum of many small contributions; and this, ultimately, is why the normal distribution is so prevalent in life.

At this point, alas, we have to conclude with a warning. In every chapter up to the present one, we dealt with exact probabilities. In this chapter, by contrast, everything is "approximately," or "roughly," or "about." In a sense, this is a necessary trade-off: as we've said, in real life there are too many factors affecting every outcome to allow for precise calculations of probabilities; if we want our subject to have applicability outside a casino, we have to put up with some imprecision.

But we should never allow ourselves to forget the ways in which we are being approximate. We say, for example, that, "given any game G, for large n the normalized bar chart of the game $G(n)$ approximates the bell-shaped curve." But what do we mean by "for large n?" How large does n have to be before this becomes a reasonable approximation, and then what sort of error of approximation are we talking about?

There are ways of answering this, if we are very careful to quantify everything and have a *lot* of mathematical tools available to us. Unfortunately, the further people get from a controlled, scientific environment, the less possible it is to fully quantify everything—and the greater the temptation grows to discuss everything in qualitative, rule-of-thumb terms. And that, ultimately, is what leads to sayings like the one attributed to Disraeli, "There are three kinds of lies: lies, damned lies, and statistics." We'll talk more about this in the concluding chapter.

14 Don't try this at home

A few years back one of us got involved, unfortunately, in a dispute between the people who run the Harvard University swimming pool and a group of the recreational swimmers who use the facility. A local swim team had been given permission to use the pool to hold some of its practices, which meant that for those hours certain lanes would not be available for recreational swimming, and some of the recreational swimmers were upset.

We did what we guess you'd expect a mathematician to do: we calculated the total number of lane hours available for recreational swimming at Harvard's facilities, and compared it to what was offered at other institutions in the area. We found Harvard had roughly three times the availability of the next most generous university, and more than 10 times as many as most others.

This did not mollify the rec swimmers. The leader of the group was particularly incensed. "Numbers!" she spat, "You can prove *anything* with *numbers!*"

So, how did numbers get such a bad reputation? In this final chapter, we'll take a look at the ways in which numbers, and in particular probability, can be misused: statistics' greatest misses, if you like.

14.1 INVERTING CONDITIONAL PROBABILITIES

One way that probability can be abused was already discussed in Chapter 9, but it bears repeating. If A and B are two events, the probability $P(A$ assuming $B)$ is not necessarily the same thing as the probability $P(B$ assuming $A)$. But as we learned recently, even we who preach on the subject don't always follow our own teachings.

One of us realized this during a visit to the doctor last month. The doctor had just received the results of a routine blood sample, which reported that the level of prostate-specific antigen (PSA) was high. The doctor then recommended a prostate biopsy. When asked why, he replied, "Well, we've found that among people with prostate cancer, a larger than usual fraction have elevated PSA."

Now, the right thing to say would clearly have been "No, you statistical illiterate! The issue is not the probability that I'd have an elevated PSA given that I had prostate cancer, but the probability that I have prostate cancer given that I have an elevated PSA! They're not the same thing, as any *Fat Chance* reader would know!" And then, surely the doctor would have appreciated a follow-up lecture on Bayes' theorem and its applications.

But, sadly, this alternate reality didn't come to pass. Instead, the patient acquiesced (as always seems to happen in the presence of doctors) and scheduled the biopsy. But it was sobering. Somewhere in the hierarchy of a health maintenance organization, someone had made a decision that, if a patient had a PSA over 4.0, he should have a biopsy. Did *that* person know Bayes' theorem, or has a decision affecting the lives of many individuals been made on the basis of a misuse of probability theory?

14.2   FALSE POSITIVES

Even if the doctor had appreciated the distinction between the probabilities $P(A$ assuming $B)$ and $P(B$ assuming $A)$, there is a secondary issue that he should have been aware of that can be a serious concern for public health. Every biopsy comes with some nonzero risk of complications, either from the procedure itself, or from follow-up surgeries that might be ordered out of an abundance of concern for an apparent abnormality that is likely to be benign. So if the likelihood that a patient with an elevated PSA actually has prostate cancer is very small, it might well be the case that a follow-up biopsy actually poses a *greater* risk to the patient's health than the risk of prostate cancer.

To illustrate with numbers (because in our view numbers make everything so much clearer), imagine a patient goes to see a doctor. The doctor performs a test with 99% reliability to screen for a particular disease—that is, 99% of people who are sick test positive and 99% of healthy people test negative. The doctor knows that only 1% of the general population has the disease. Now the question is: if the patient tests positive, what are the chances the patient has the disease?

Since the test is 99% reliable you might think that the answer is 99% but no: this number is the probability that the patient receives a positive test result assuming they are sick. Instead, we're asked to compute the probability the patient is sick assuming a positive test result, which, as we're trying to emphasize, is generally not the same thing. Since the probability of being sick is $1/100$, we know from our investigation into conditional probabilities in Section 9.2 that

$$P(\text{positive and sick}) = P(\text{sick}) \cdot P(\text{positive assuming sick}) = \frac{1}{100} \cdot \frac{99}{100} = \frac{99}{10^4}.$$

To apply Bayes' theorem, we need to compute the overall probability of a positive test result, which is the sum of the probability of getting a positive result while being sick and the probability of getting a positive result while being well. The latter we can also compute by multiplying:

$$P(\text{positive and well}) = P(\text{well}) \cdot P(\text{positive assuming well}) = \frac{99}{100} \cdot \frac{11}{100} = \frac{99}{10^4}.$$

So

$$P(\text{positive}) = P(\text{positive and sick}) + P(\text{positive and well}) = \frac{198}{10^4}.$$

Finally, by Bayes' theorem, the probability of being sick assuming a positive result is computed by diving the probability of being sick and having a positive test result by the probability of having a positive result. Thus

$$P(\text{sick assuming positive}) = \frac{P(\text{sick and positive})}{P(\text{positive})} = \frac{\frac{99}{10^4}}{\frac{198}{10^4}} = \frac{1}{2}.$$

So despite the 99% accuracy of the test, a patient who receive a positive test result has only a 50% chance of actually having the disease in question!

Note the probability *not* having the disease assuming a negative test result is much higher than 50% since any individual is much more likely not to have the disease in any case and not being sick correlates positively with having a negative test result. (We realize this sentence is a mouthful. One exercise for the reader is to figure out what on earth we're trying to say. A second exercise is to compute the probability a patient being well assuming a negative test result.) This positive correlation means that the probability of not being sick if you get a negative test result is even larger than the probability of not being sick, which already is 99%.

For this reason, public health officials typically advise doctors *against* screening general populations for sufficiently rare diseases because a positive test result is highly likely to be a false positive. For instance, unless there is a strong family history or other contributing factors, doctors do not recommend women in their twenties or thirties to get mammograms to test for breast cancer. And a population-wide screening of infants for neuroblastoma in Japan was discontinued in 2004 as the problem of false positives began to be recognized.

14.3 ABUSE OF AVERAGES

After that heavy topic, let's move back to the world of fantasy! Let's say that Tracy and Paul have abandoned politics entirely, and have decided instead to become professional baseball players.

Now, baseball is a statistics-hungry sport, and probably the king of all baseball statistics is the batting average. This is essentially the number of times a batter has gotten a hit, divided by the total number of times they've been at bat. (As such, it's a number between 0 and 1, but it's usually written as a three-digit number, with the decimal point in front omitted. To be more precise, we'll keep the decimal place but round to the nearest thousandth as per convention.)

Next, suppose that Tracy's and Paul's careers as baseball players span exactly the same range of years. Suppose moreover that in every one of these years, Tracy's batting average was greater than Paul's. It stands to reason that Tracy's lifetime batting average would be higher than Paul's, right?

No, in fact it doesn't. It's possible for Tracy to have a higher batting average than Paul every single year that they played, but for Paul's career batting average to be greater! Here's an example, based on a playing career of two years:

year	T at-bats	T hits	T average	P at-bats	P hits	P average
2009	10	4	.400	100	35	.350
2010	100	25	.250	10	2	.200
totals	110	29	.264	110	37	.318

Again: in each year, Tracy's average was higher than Paul's (.400 versus .350; .250 versus .200); but Paul's lifetime average is .318, which is higher than Tracy's .264.

This is an example of what's called *Simpson's paradox*. It's not hard to spot what's going on, once you think about it: if you look at the table, you'll see that the great majority of Paul's lifetime at-bats occurred in 2009, a year in which both did quite well. A majority of Tracy's at-bats, by contrast, occurred in 2010, a year in which they both struggled.

Now, this may be a fantastical, made-up situation (though not for the reason you might be thinking—there's talk that someday soon a woman will be drafted by Major League Baseball), but Simpson's paradox has occurred in serious settings in real life as well. A famous example involved admissions to the UC Berkeley graduate school. It was found that, among all applicants to UC Berkeley graduate programs, the admissions rate (the numbers of admits divided by the number of applicants) was higher among men than among women; this was taken as clear indication of gender bias. But when the figures were broken down further, it was found that *in every single department, the admissions rate was higher among women than among men.*

How was this possible? Basically, it's the same story as with Tracy and Paul's batting averages: different departments had drastically different overall rates of admission, and more women applied to the departments with lower admissions rates.

14.4   RANDOM CORRELATION

Say you flip a pair of coins. The probability that they agree—either both come up heads, or both come up tails—is exactly one-half. Now suppose that you were to flip this pair of coins, say, 100 times. What's the likelihood that they would agree 65 or more times?

We know how to do this; without dragging you through the process one more time, the answer is that the probability of this occurring randomly is just 0.00135, or roughly one in a thousand. If this actually occurred, you'd certainly be justified in thinking that there was a connection between the two coins—that one influenced the other (if that weren't patently ridiculous: entangled coins don't exist as far as we know).

Now suppose you repeat this experiment 5,000 times, with 5,000 different pairs of coins. That is, we're asking you to take 5,000 pairs of coins, flip each pair 100 times, and see how often each pair of coins agreed. If the probability that two coins agree 65 or more times out of 100 is just 0.00135, the probability that they agree 64 or fewer times is 0.99865. Now the chances the *none* of the 5,000 pairs of coins agrees more than 65 times is

$$(0.99865)^{5000} \sim 0.0012.$$

In other words, the shoe's on the other foot: it's almost certain that at least one of the 5,000 pairs of coins will display exactly this spookily entangled behavior. But we wouldn't deduce from this that one particular pair of coins had a mysterious, mystical connection; we'd just chalk it up to the fact that even highly unlikely events will occur occasionally, if you repeat the experiment enough times.

Next, imagine that we take 100 coins, and flip them 100 times. If you think about it, among the 100 coins there are

$$\binom{100}{2} = \frac{100 \cdot 99}{2} = 4,950$$

pairs. So this is like the last experiment: you'd expect that some pair of the 100 coins would agree at least 65 of the 100 times. Again, though, you wouldn't deduce a connection between those two particular coins.

Finally, suppose you're a medical researcher, investigating correlations between various risk factors for various diseases. You prepare a lengthy questionnaire, consisting of 100 questions; for simplicity, assume for the moment these are all yes-or-no questions, to which you anticipate that roughly half the respondents will answer "yes" and half "no." You assemble 100 subjects who are willing to take the time to fill out your questionnaire; they do so, and you examine the results.

Now, if you were to look at any two of the questions and find that more than 65 of the 100 respondents had responded the same way, you'd be justified in at least considering the possibility that there was a correlation (if not a causal relationship) between these two factors; as we said, the odds of this happening if the two were indeed unrelated is just one in a thousand. But you're not looking at the answers to just two questions; you're looking at 100 questions. And this is very much much like the last example, in which you flipped 100 coins: even if all the questions in your survey dealt with absolutely unrelated factors—were you born on an odd-numbered day of the month, or an even-numbered one? Does your first name begin with a letter between A and K, or between L and Z?—*you would be almost guaranteed to observe a connection between the answers to some pair of these questions.*

This is what we may call *random correlation*. It's the phenomenon that if you look at enough data, you can and almost certainly will observe an apparent correlation between two factors. Of course, having found this apparent connection, the scientific method then dictates that you do a follow-up study on just those two factors; if you ask, say, 1,000 people those two questions and find more than 65% agreement, then you can reasonably deduce a connection (but more likely—as in the case where the questions are truly unrelated—the apparent correlation will not be borne out).

Unfortunately, the scientific method is not always observed in these matters. Let's face it: the scientific method can be a pain. It requires follow-up studies, which take additional time, money, and resources. And—especially when the results have political or economic implications, and the apparent results of the initial test are to your liking—there's a temptation to take the initial results and just run with them. But as the example here suggests, that leads in the end to garbage that gives numbers a bad name.

14.5   WHAT WE THINK IS RANDOM, ISN'T; WHAT WE THINK ISN'T RANDOM, IS

The human brain is a funny thing, as anyone who's experienced puberty can tell you. For example, the concept of number, and the ability to manipulate them, seems almost to be hardwired into us; numbers have been around almost as long as language. Likewise, the notion of probability seems natural to us; gambling has been around almost as long as either.

But that doesn't mean we're good at it! In this section we want to describe one of the ways in which our sense of probability, and of randomness in particular, can and does lead us astray.

Before we get into it, let's do an experiment. First, get a pencil and paper and write down a string of 100 letters chosen at random between *H*s and *T*s—in other words, what you'd expect to see if you took out a fair coin, flipped it 100 times, and wrote down the sequence of resulting Hs and Ts. Second, take out a fair coin, flip it 100 times, and write down the resulting sequence of Hs and Ts.

You now have two sequences of 100 letters, chosen between Hs and Ts. Do they look alike? In fact, if you're like most people, they'll differ in a number of respects; a trained statistician or probabilist would be able to tell right away which is the sequence you made up, and which is the one resulting from coin flips. We're going to describe here one of the ways: the presence, or absence of *streaks*.

To set this up, let's pose a problem in probability: if we flip a fair coin 100 times in sequence, *how long a streak of heads would we expect to see?* We can express this question in terms of a game, as follows: in one iteration of the game, we flip a coin 100 times and the payoff is n dollars, where n is the length of the longest streak of heads that occurs. The payoff could thus be as high as 100 (if all flips come up heads) or as low as 1 (if, for example, the sequence simply alternates between heads and tails) or even 0 (if they're all tails). But these outcomes are all highly unlikely; we'd like to know what the most likely outcomes are. For example, we can ask: what's the expected value of this game?

We're not going to solve this problem here. It's like the gambler's ruin problem we discussed in Section 10.3, in that it requires us to consider a more general form of the problem (any number of flips, a possibly unfair coin) and apply conditional probability; and in any event it calls for way more algebraic techniques than we have. But we can give you the answer: either by slogging through the algebra, or simply asking a computer to play the game a large number of times and tracking the outcomes, we can see that on average the longest streak of heads will be 6 in a row.

Now go back to those two sequences of Hs and Ts you came up with, and find the longest sequence of Hs in each. By what we just said, in the truly random sequence—the one you generated by actually flipping a coin 100 times—there is apt to be a streak of 6 or more heads. What about the sequence you made up yourself, supposedly at random? Well, if you're like most people, the longest streak of heads is likely to be 5 heads or shorter. In other words, the human brain tends to discount the possibility that apparently non-random phenomena, like streaks, may in fact be simply random.

We can debate the evolutionary reasons why this might be so. An important survival skill for humans is the ability to distinguish random events from causal ones: the random ones we can't do anything about, but causal events may require action on our part. In that setting, mistaking a random phenomenon for a causal one is relatively harmless: thinking that the weather is bad just because you've somehow pissed off the weather gods, and offering them a burnt sacrifice, may not do much good, but doesn't do any harm, either (except, of course, for the creature sacrificed). But the opposite error—failing to take action to deal with a problem that you can in fact affect, because you attribute it to randomness—can be deadly. (NOT getting into climate change, here.) In that respect, it makes sense that we're hardwired to underestimate the likelihood of phenomena like streaks.

The vagaries of our mental processes aside, the fact is that as a species we are not all that great at distinguishing random phenomena. One famous instance of this is what's called the *hot hand fallacy* in basketball. Basically, this is the belief—held, according to one survey, by over 90% of basketball fans—that a player can get "hot" or "cold;" and that a player who's hot is more likely to make their next shot than their career average would predict, and vice versa. (Interestingly, many people also subscribe to the opposite view, called the *gambler's fallacy*, that a coin that has come up heads several times in a row is "due" to come up tails, so that a tail on the next flip is more likely than it would be in general.)

According to the hot hand fallacy, if we were to look at the results of successive shots taken by a basketball player with a career average of, say, 50%, you would see more and longer streaks than you would if you examined the results of that many coin flips. In a famous paper that has led to three decades of debate, Amos Tversky, Thomas Gilovich and Robert Vallone examined the statistics and found no evidence for this. This has led to decades of research, not all of which has sided with Tversky, Gilovich and Vallone; we're not going to take sides here, but we did want to make the point that our sense of probability is often subject to unconscious biases, and that we should take it with a grain of salt. That is in part why, throughout this book, we've urged you to think about each probability problem we've proposed, and write down your guess, before actually working out the answer: to identify the ways in which your intuition may lead you astray.

A Boxed formulas

The number of whole numbers between k and n inclusive is
$$n - k + 1.$$
(p. 5)

The product
$$n \cdot (n - 1) \cdot (n - 2) \cdots 3 \cdot 2 \cdot 1$$
of the numbers from 1 to n is written $n!$ and called "n factorial."
(p. 17)

The number of ways of making a sequence of independent choices is the product of the number of choices at each step.
(p. 12)

The number of sequences of k objects chosen without repetition from a collection of n objects is
$$\frac{n!}{(n - k)!}.$$
(p. 18)

The number of sequences of k objects chosen from a collection of n objects is
$$n^k.$$
(p. 15)

The number of objects in a collection that satisfy some condition is equal to the total number of objects in the collection minus the number of those that don't.
(p. 21)

The number of sequences of k objects chosen without repetition from a collection of n objects is
$$n \cdot (n - 1) \cdot (n - 2) \cdots (n - k + 1).$$
(p. 16)

For any two sets of elements in a given pool, the number of elements in their union is equal to the sum of the number of elements in each set minus the number of elements in their intersection:
$$\#(A \cup B) = \#A + \#B - \#(A \cap B).$$
(p. 27)

The number of ways of choosing a collection of k objects, without repetition, from among n objects is

$$\frac{n!}{k!\,(n-k)!}.$$

(p. 33)

The number of ways of choosing a collection of k objects from among n objects, with repetitions allowed, is

$$\binom{n+k-1}{k} = \frac{(n+k-1)!}{k!\,(n-1)!}.$$

(p. 79)

The number of ways of distributing n objects into groups of size $a_1, a_2, \ldots, a_k$ is

$$\frac{n!}{a_1!\cdot a_2!\cdot \cdots \cdot a_k!}.$$

(p. 43)

The nth Catalan number may be computed from the previous Catalan numbers by:

$$c_n = c_0 c_{n-1} + c_1 c_{n-2} + c_2 c_{n-3}$$
$$+ \cdots + c_{n-2}c_1 + c_{n-1}c_0.$$

(p. 83)

The probability of getting exactly k heads in n flips is

$$\frac{\binom{n}{k}}{2^n}.$$

(p. 48)

The nth Catalan number c_n is given by the formula

$$c_n = \frac{1}{n+1}\binom{2n}{n}.$$

(p. 85)

In Pascal's triangle, each binomial coefficient s the sum of the two above it:

$$\binom{n}{k} = \binom{n-1}{k-1} + \binom{n-1}{k}.$$

(p. 68)

The *expected value* of a game is the average payoff per play.

(p. 97)

The coefficient of $x^k y^{n-k}$ in $(x+y)^n$ is

$$\binom{n}{k}.$$

(p. 73)

If an experiment has k possible results, occurring with probabilities $p_1, \ldots, p_k$ and payoffs $a_1, \ldots, a_k$, the *expected value* of the experiment is

$$\text{ev} = p_1 a_1 + p_2 a_2 + \cdots + p_k a_k.$$

(p. 102)

In a situation where either of two events A or B occurs, but not both, and W is a third event, whose outcome may depend on A or B:

$$P(W) = P(A) \cdot P(W \text{ assuming } A)$$
$$+ P(B) \cdot P(W \text{ assuming } B).$$

(p. 112)

In n trials of an event where the outcome A occurs with probability p, the probability of having A occur exactly k times is:

$P(\text{exactly } k \text{ As})$
$$= \binom{n}{k} p^k (1 - p)^{n-k}.$$

(p. 137)

Given two events, the first of which has two possible outcomes A or B, and the second of which has two possible outcomes W or L, if

$$P(W) = P(W \text{ assuming } A)$$
$$= P(W \text{ assuming } B),$$

then the two events are *independent*.

(p. 116)

The *variance* of a game is the average of the square of the differences between the payoffs and the expected value, weighted according to their probability:

$\text{var}(G)$
$$= p_1(a_1 - \text{ev})^2 + \cdots + p_k(a_k - \text{ev})^2.$$

(p. 161)

Bayes' theorem: Given one event with outcomes A and B and a second event with outcomes M and N:

$P(A \text{ and } M)$
$$= P(A) \cdot P(M \text{ assuming } A)$$
$$= P(M) \cdot P(A \text{ assuming } M).$$

Consequently:

$P(M \text{ assuming } A)$
$$= P(A \text{ assuming } M) \cdot \frac{P(M)}{P(A)}.$$

(p. 123)

For any game G, its *normalized form* is the game:

$$G_0 = \frac{G - \text{ev}}{\sqrt{\text{var}}},$$

which has the same outcomes and probabilities but with payoffs adjusted so that the expected value is 0 and the variance is 1.

(p. 165)

B Normal table

**Probability Content
from -oo to Z**

z	0.00	0.01	0.02	0.03	0.04	0.05	0.06	0.07	0.08	0.09
0.0	0.5000	0.5040	0.5080	0.5120	0.5160	0.5199	0.5239	0.5279	0.5319	0.5359
0.1	0.5398	0.5438	0.5478	0.5517	0.5557	0.5596	0.5636	0.5675	0.5714	0.5753
0.2	0.5793	0.5832	0.5871	0.5910	0.5948	0.5987	0.6026	0.6064	0.6103	0.6141
0.3	0.6179	0.6217	0.6255	0.6293	0.6331	0.6368	0.6406	0.6443	0.6480	0.6517
0.4	0.6554	0.6591	0.6628	0.6664	0.6700	0.6736	0.6772	0.6808	0.6844	0.6879
0.5	0.6915	0.6950	0.6985	0.7019	0.7054	0.7088	0.7123	0.7157	0.7190	0.7224
0.6	0.7257	0.7291	0.7324	0.7357	0.7389	0.7422	0.7454	0.7486	0.7517	0.7549
0.7	0.7580	0.7611	0.7642	0.7673	0.7704	0.7734	0.7764	0.7794	0.7823	0.7852
0.8	0.7881	0.7910	0.7939	0.7967	0.7995	0.8023	0.8051	0.8078	0.8106	0.8133
0.9	0.8159	0.8186	0.8212	0.8238	0.8264	0.8289	0.8315	0.8340	0.8365	0.8389
1.0	0.8413	0.8438	0.8461	0.8485	0.8508	0.8531	0.8554	0.8577	0.8599	0.8621
1.1	0.8643	0.8665	0.8686	0.8708	0.8729	0.8749	0.8770	0.8790	0.8810	0.8830
1.2	0.8849	0.8869	0.8888	0.8907	0.8925	0.8944	0.8962	0.8980	0.8997	0.9015
1.3	0.9032	0.9049	0.9066	0.9082	0.9099	0.9115	0.9131	0.9147	0.9162	0.9177
1.4	0.9192	0.9207	0.9222	0.9236	0.9251	0.9265	0.9279	0.9292	0.9306	0.9319
1.5	0.9332	0.9345	0.9357	0.9370	0.9382	0.9394	0.9406	0.9418	0.9429	0.9441
1.6	0.9452	0.9463	0.9474	0.9484	0.9495	0.9505	0.9515	0.9525	0.9535	0.9545
1.7	0.9554	0.9564	0.9573	0.9582	0.9591	0.9599	0.9608	0.9616	0.9625	0.9633
1.8	0.9641	0.9649	0.9656	0.9664	0.9671	0.9678	0.9686	0.9693	0.9699	0.9706
1.9	0.9713	0.9719	0.9726	0.9732	0.9738	0.9744	0.9750	0.9756	0.9761	0.9767
2.0	0.9772	0.9778	0.9783	0.9788	0.9793	0.9798	0.9803	0.9808	0.9812	0.9817
2.1	0.9821	0.9826	0.9830	0.9834	0.9838	0.9842	0.9846	0.9850	0.9854	0.9857
2.2	0.9861	0.9864	0.9868	0.9871	0.9875	0.9878	0.9881	0.9884	0.9887	0.9890
2.3	0.9893	0.9896	0.9898	0.9901	0.9904	0.9906	0.9909	0.9911	0.9913	0.9916
2.4	0.9918	0.9920	0.9922	0.9925	0.9927	0.9929	0.9931	0.9932	0.9934	0.9936
2.5	0.9938	0.9940	0.9941	0.9943	0.9945	0.9946	0.9948	0.9949	0.9951	0.9952
2.6	0.9953	0.9955	0.9956	0.9957	0.9959	0.9960	0.9961	0.9962	0.9963	0.9964
2.7	0.9965	0.9966	0.9967	0.9968	0.9969	0.9970	0.9971	0.9972	0.9973	0.9974
2.8	0.9974	0.9975	0.9976	0.9977	0.9977	0.9978	0.9979	0.9979	0.9980	0.9981
2.9	0.9981	0.9982	0.9982	0.9983	0.9984	0.9984	0.9985	0.9985	0.9986	0.9986
3.0	0.9987	0.9987	0.9987	0.9988	0.9988	0.9989	0.9989	0.9989	0.9990	0.9990

Far Right Tail Probabilities

Z	P{Z to ∞}	Z	P{Z to ∞}	Z	P{Z to ∞}	Z	P{Z to ∞}
2.0	0.02275	3.0	0.001350	4.0	0.00003167	5.0	2.867 E-7
2.1	0.01786	3.1	0.0009676	4.1	0.00002066	5.5	1.899 E-8
2.2	0.01390	3.2	0.0006871	4.2	0.00001335	6.0	9.866 E-10
2.3	0.01072	3.3	0.0004834	4.3	0.00000854	6.5	4.016 E-11
2.4	0.00820	3.4	0.0003369	4.4	0.000005413	7.0	1.280 E-12
2.5	0.00621	3.5	0.0002326	4.5	0.000003398	7.5	3.191 E-14
2.6	0.004661	3.6	0.0001591	4.6	0.000002112	8.0	6.221 E-16
2.7	0.003467	3.7	0.0001078	4.7	0.000001300	8.5	9.480 E-18
2.8	0.002555	3.8	0.00007235	4.8	7.933 E-7	9.0	1.129 E-19
2.9	0.001866	3.9	0.00004810	4.9	4.792 E-7	9.5	1.049 E-21

Index